AIRLIFT!

AIRLIFT!

The Story of the Military Airlift Command

Marcella Thum and Gladys Thum

Illustrated with photographs

DODD, MEAD & COMPANY   New York

Distributed in Canada by
McClelland and Stewart Limited, Toronto
Manufactured in the United States of America
1   2   3   4   5   6   7   8   9   10

Library of Congress Cataloging-in-Publication Data

Thum, Marcella.
    Airlift! : the story of the Military Airlift Command.

    Includes index.
    Summary: A history of MAC, a military organization
which is the "backbone of deterrence" for United States
fighting forces, the largest peacetime cargo airline
in the world, and a humanitarian airlift in times of
disaster.
    1. United States.   Air Force.   Military Airlift
Command—History—Juvenile literature.   [1. United States.
Air Force.   Military Airlift Command—History]   I. Thum,
Gladys.   II. Title.
UG633.T46   1986   358.4′4′0973   85–27397
ISBN 0–396–08529–6

In tribute to the Airlifters of the United States Air Force, who valiantly served and gallantly died in support of this nation's dedication to the principles of liberty and freedom.

ACKNOWLEDGMENTS

Grateful recognition is given to Lt. Col. Edward Wittel, Commander, Det 1, 1361st Aerospace Audiovisual Service; John Fuller, Donald Little, Joylyn Gustin, Hq MAC Historians; Betty Kennedy, 375th AAW Historian; Lt. Col. Raymond Jojola, Hq MAC Weather; David Wilson, Darlene Fuller, Linda Frierdich, and Maj. Portia McCracken, Hq MAC Public Affairs; Lt. Tom Dolney, 375th AAW Public Affairs; Maj. Byron Howard and Tom O'Laughlin of *The MAC Flyer*; Maj. Thornton Phillips, Editor, *Airlift*; and Vivian White of the Research Center, USAF Museum, Dayton, Ohio.

Special thanks to Lt. Cols. Billie Carpenter and John Johanek, present and past commandants, Airlift Operations School, and the following faculty and staff of the School: Lt. Col. David Myers, Lt. Col. Clement Wehner, Maj. Carol Henry, Maj. Ronny Smith, Maj. Peter Nelson, Maj. Danny Dees, and Maj. Kent Douglas—for their unfailing assistance and support.

Contents

FOREWORD

From its beginning in 1941 as the Air Corps Ferrying Command, through three wars and innumerable humanitarian missions, the Military Airlift Command has served the United States at home and abroad by providing airlift, when and where needed, anywhere in the world. Although its primary mission is airlift in support of national strategy and national policy, MAC also has many other varied and important responsibilities: aeromedical, special air mission, operational support airlift, combat rescue, special operations, audiovisual and weather services.

The authors of this book have not attempted to write a complete, in-depth presentation of the U.S. Air Force's Military Airlift Command. They have, however, brought together little-known information on the proud history of the oldest command in the USAF, as well as showing what MAC is and does.

As a former Commander in Chief of the Military Airlift Command, I believe the importance of military airlift in the past and present should be better known; even more important is increased awareness of the vital role the Military Airlift Command will play in the future.

General William G. Moore, Jr.
USAF (Ret.)

CHAPTER 1

"You Call, We Haul"

By the spring of 1941, during World War II, England was reeling under Nazi air attacks and desperately in need of more bombers and fighter planes. Although America was still officially neutral, President Franklin D. Roosevelt on May 28, 1941, ordered the Secretary of War to speed up the delivery of American-built bombers to England. The very next day the Air Corps Ferrying Command was activated. The Command was not only the beginning of military airlift—the transporting of planes, troops, and war material by air—but the predecessor of the Military Airlift Command, the oldest command in the United States Air Force.

The original mission of the new Command was to ferry American-built British lend-lease airplanes from factories in the United States to departure points in Canada for transport by British-Canadian pilots to Britain. Soon, however, the Ferrying Command also became a military air transport service for the War Department. Military passengers and civilians on diplomatic missions were ferried aboard B-24 Liberator bombers from Bolling Field in Washington, D.C., to Montreal and Newfoundland, then via the North Atlantic to Scotland. Often the only seating available for the passengers aboard the B-24s was in the bomb bays!

Despite the fact that few American pilots and crews prior to 1941 had any experience in flying over water or at night, within six months the Ferrying Command, under the command of Colonel Robert Olds, had delivered 1,350 aircraft to the East Coast, flying night and day. With the Japanese attack at Pearl Harbor on December 7, 1941, and America's entrance into World War II, the Ferrying Command began its much-expanded foreign ferrying operations by delivering four B-24 Liberator bombers to the Middle East. The Ferrying Command also started mapping and charting air routes that had never been traveled by aircraft before. Eventually these new air routes would circle the globe.

In June, 1942, the Ferrying Command was renamed the Air Transport Command (ATC). The mission of the ATC was not much different from the Ferrying Command, but greatly expanded. ATC handled airlift not just for the Army Air Forces but for the entire War Department. And by August, 1942, ATC took on the important new mission of air-evacuating sick and wounded American servicemen and women throughout the world. It was a large order, considering that the new Command had less than a dozen transport planes and even fewer airfields with which to perform its duties.

Factories were working day and night turning out fighter planes and bombers for the war effort. At first, ATC had to purchase

Boeing clippers, Martin Flying Boats, and Boeing Stratoliners (C-75s) from commercial airlines to use as transport planes. But lack of aircraft was only one of the new command's problems.

The laborious mapping of new, previously uncharted air routes across the Atlantic and Pacific, which are as essential to airlift as highways are to automobiles, had only just begun when war broke out in the Pacific. Japan's occupation of many of the Pacific islands —the Philippines, Wake, Guadalcanal, and Midway—meant that new air routes had to be charted and new island airfields built before desperately needed men and supplies could be airlifted to the Pacific war zone.

The most pressing problem facing ATC, though, was the lack of trained pilots and crews to fly their transport planes. Most of the experienced pilots and crews were already flying fighters and bombers. Once again, ATC turned to the commercial airlines. Civilian pilots and crews were inducted into the service and civilian airlines placed under contract to the Army Air Force. Even with help from the commercial airlines, however, there were still never enough pilots to man the transport planes.

Then one day Col. William H. Tunner, head of the ATC ferrying division's domestic wing, discovered that the wife of one of his officers was a pilot, and exclaimed, "I'm combing the woods for pilots, and here's one right under my nose. Are there many more women like your wife?"

"Why don't you ask her?" Major Love replied.

Although there was considerable opposition to the idea of women pilots flying for the military—so much so that the women pilots were not granted military status and only flew within the United States—Nancy Love managed to start the Women's Auxiliary Ferrying Squadron (WAFS). Jacqueline Cochran later incorporated the organization into the Women's Air Force Service Pilots (WASP) program.

At first, the women pilots ferried small training planes. Soon, however, they were ferrying the Army Air Force's biggest bombers and high-powered fighter planes. They also acted as test pilots and towed targets for antiaircraft artillery practice. By 1944, half of all the pilots ferrying fighter planes were women, and three-fourths of all domestic deliveries of America's military planes of all types were accomplished with WASPs in the cockpit. More than one thousand women wore the wings of Women's Air Force Service Pilots.

While women pilots were breaking new ground in America, the greatest sustained, intensive use of airlift in history was being undertaken on the other side of the world. In 1942, Japan blocked all water and land access to China, effectively cutting the supply line to Chinese and American troops fighting the Japanese in China. The only remaining lifeline into China was by air from Eastern India over the awesome Himalaya Mountains into the Yunnan Province of China. This 500-mile air route was known as "the Hump."

The task of flying "the Hump," bringing vitally needed guns and supplies to Chinese and American troops in China, was given to the India-China Wing of the Air Transport Command. Flying "the Hump" became one of the epics of airlift history.

The uncharted Himalaya Mountains soared from 14,000 to 16,500 feet. Many of the peaks were shrouded in constant cloud cover. The weather was treacherous, from torrential rainstorms, called monsoons, to violent turbulence that could cause a plane to plummet 3,000 feet a minute, and freezing temperatures that

16

Four members of the Women's Air Force Service Pilots (WASPs) just after graduation from B-17 school at Lockbourne Field, Ohio, 1944.

could cause severe icing on the aircraft. Radio beams were almost unknown. When planes went down in the Himalayas, few of the crews were ever seen again.

Pilots who avoided the higher peaks to the north, and took the southern route over the Himalayas, had to cross over northern Burma into China and fared little better. Burma was Japanese-held territory and ATC's unarmed C-46 Commandos and C-47 Gooney Birds, and later, the C-54 Skymasters, came under attack from Japanese Zeros. Although often shot at, transport planes were not

17

built or equipped to shoot back. If shot down over Burma, the dense jungle soon obliterated any trace of the planes, and any survivors had to fend off tribes of headhunters who lived in the area.

Some ingenious pilots of C-87s (converted B-24s) also flying "the Hump," placed black-painted bamboo poles, the thickness of gun barrels, into their ships' noses, sides, and tails, hoping their planes would resemble Liberator bombers and frighten off the Zeros!

A C-46 Commando of the Air Transport Command flying "the Hump," the snow-capped Himalayas between India and China, in World War II.

Nevertheless, despite shortages of gasoline and airplane parts, primitive living conditions, and crews suffering from malaria and dysentery, and three American lives lost for every thousand tons flown into China, "the Hump" airlift continued. For the first time in military history, airlift kept an entire combat theater alive with "beans and bullets." One historian has said, "Together with Hannibal's crossing of the Alps, the Hump operation will go down in the annals of military history as one of the most difficult logistics missions accomplished by any military force."

While the Air Transport Command was the beginning of strategic airlift—transporting planes, troops, and supplies from the United States to a war zone—in Europe the IX Troop Carrier Command was the beginning of tactical airlift—transporting troops and supplies directly into battle. The Troop Carrier Command, which was not part of ATC, used C-47s and gliders to carry airborne troops directly into battle in Nazi-held France on D-Day.

Wars throughout history have always been won or lost not just by which side is the strongest, but by logistics—which side can supply and resupply their army the fastest with the most troops and equipment. Surface and sealift transportation are still essential in any long-term conflict. But for the first time during World War II, troops and war materials were transported to armies in the field with a speed and efficiency and over distances never before possible. Airlift had added a new dimension to warfare.

After the end of World War II, the Army Air Corps became a separate military service, the United States Air Force. In 1948, the Air Transport Command was renamed the Military Air Transport Service (MATS) and became a major command under USAF, operating a global air transport system for the Department of Defense. In addition to its strategic airlift mission, aeromedical evacuation, and transporting important dignitaries, such as the President of the United States, MATS became responsible for Air Weather, Air Rescue services and Airways and Air Communications Service. However, Tactical Air Command, a separate command from MATS, became responsible, for the most part, for tactical airlift, transporting airborne troops and their equipment into forward combat areas.

Despite its vital missions, MATS was cut back severely after the war, as were other military services. So much so that when the Russians suddenly threw their blockade around West Berlin, the C-47 transports of MATS, the workhorses of World War II, were almost worn out.

At the close of World War II, West Berlin was an island surrounded by the Russian zone of occupied Germany. Berlin had been split in two. The Western powers—United States, Great Britain, and France—controlled West Berlin; the Russians, East Berlin. In June, 1948, Russia blockaded all road and water access to West Berlin. Without food, medical supplies, and coal vital to lighting and heating homes and factories, the Communists hoped to force the Western powers to abandon West Berlin. Russian tanks were massing on the German borders. Many believed the blockade could be the start of World War III.

Gen. Lucius D. Clay, the U.S. military governor of West Berlin, telephoned Maj. Gen. Curtis LeMay, commander of U.S. Air Forces in Europe, and asked an urgent question, "Can you haul coal?"

General LeMay replied with what has become the traditional rallying cry of airlifters everywhere, "General, we can haul anything."

The job of running the Berlin airlift was given to the former head of ATC's ferrying division, domestic wing, now Maj. Gen. William H. Tunner, Deputy Commander, Air Transport, MATS. General Tunner was put in charge of the 1st Airlift Task Force in Berlin and saw at once that the larger C-54 Skymasters were necessary to take over the job of supplying the besieged city.

Even with the larger C-54s, there were almost insurmountable problems facing the Berlin airlift, or "Operation Vittles" as it was called. Tempelhof Airport in the middle of Berlin is girdled by hills and apartment buildings, making air traffic difficult under the best of conditions. In the winter, low-lying fogs and sleet storms made takeoffs and landings even more dangerous. Yet to keep the city alive, each day hundreds of American and British aircraft had to land, off-load their valuable cargo, and take off again from the airport. Instead, because of the heavy traffic, airplanes were "stacked up" over the airfield, wasting valuable gasoline and time, waiting to land.

General Tunner's staff set up a new traffic pattern. Each loaded aircraft arriving at Tempelhof was spaced exactly three minutes apart, flying at 200 miles an hour. To maintain that traffic pattern, all planes were required to pass checkpoints at a precise height and at an exact time. Each plane then made a straightaway pass at the runway and if the pilot failed to make the landing, he was ordered to climb and return to his home base from which he would make a fresh start.

The pilots didn't even break their rigidly controlled traffic pattern when Russian fighter planes dived in front of the transport airplanes, and shells and paratroopers were dropped close to the C-54s. In spite of any and all obstacles, the Berlin airlift continued, a remarkable 620 round-trip flights a day. The airlift brought everything from coal and flour and medical supplies to milk and candy

LAST VITTLES FLIGHT
1783 5727 TONS AIRLIFTED
TO BERLIN

UNITED STATES AIR FOR

The markings on this MATS C-54 tell the story: "Last Vittles flight airlifted to Berlin."

for the children of West Berlin. In May, 1949, the Russians finally called off their blockade. The Berlin airlift continued till September, by which time American and British air transports had delivered 2.5 million tons of material to Berlin.

The valuable experience MATS gained in flying the Berlin airlift came in handy a year later when North Korea invaded South Korea, and America was once again at war. A MATS C-54 was the first plane destroyed in that conflict.

The Pacific airlift supporting the United Nations forces in Korea was one of the longest aerial supply lines in history, a logistical pipeline of nearly 10,000 nautical miles. The shortest route from the West Coast of the United States to the combat zone required 30 hours of flying time. It took six agonizing weeks to move two Army divisions from their bases in the United States to the Korean front.

Once again, as in World War II, because of a shortage of planes and personnel, MATS had to turn to civilian airlines to supply additional airlift to the war zone. In 1952, President Harry Truman established the Civil Reserve Air Fleet (CRAF), which is still in existence today. With CRAF, a planned use of civil carriers, more than 300 commercial aircraft can be added to U.S. airlift capability in times of national emergency or war.

CRAF nearly doubled the U.S. long-range airlift capability in Korea. Each day more than 100 tons of emergency military items were flown from the United States to Japan for transshipment to the fighting units in Korea.

On the battlefields in Korea, wherever troops were entrenched, from rice paddies to mountaintops, airlift provided guns and ammunition, food and medicine. MATS Air Rescue helicopters often penetrated 125 miles behind enemy lines to rescue downed United Nations crewmembers, and for those troops wounded in action, airlift meant medical care and a hospital bed, often within twenty-four hours.

After the Korean War, various "brush-fire" wars and political crises erupted around the world in the 1950s and '60s. The Suez, Lebanon, the Congo, the Dominican Republic, and Pakistan were among some of the trouble spots where violence suddenly exploded. A fast-reacting, highly mobile force was needed to deter, contain, or end conflicts posing a threat to the United States. Modern jet airlift gave MATS the ability to position troops and supplies when and where they were needed.

It wasn't until the Vietnam War, however, that a jet aircraft, the C-141 Starlifter, was specially designed, engineered, and built to meet military standards as a troop and cargo carrier. The new C-141, along with the older C-133 Cargomaster, airlifted 10,355 men of the 101st Airborne Division and 5,118 tons of equipment from Fort Campbell, Kentucky, to Bien Hoa in South Vietnam in 42 days. Unloading operations of each plane required twenty minutes at the maximum in the largest and longest strategic military airlift ever attempted to a combat zone from the United States.

Again, as in Korea, the MATS Air Rescue helicopters swooped down out of the sky to rescue pilots downed behind enemy lines. By the end of the war, they had saved 4,120 human beings from death, suffering, or captivity.

Tactical airlift, which at that time was still under the Tactical Air Command, airlifted supplies directly into the front lines, such as the besieged U.S. Marine base at Khe Sanh. Under constant enemy fire from machine guns and mortars, C-130s continued to unload food and ammunition, airdropping the supplies when the runway was destroyed, in the biggest single parachute airdrop operation in U.S. military history.

In Vietnam, airlifters called themselves "trash haulers." It was a modest term to describe the courage it took to fly on many of their missions. One airlift transport pilot described the difference between flying the C-130 and the F-4 fighter plane. "The F-4 pilot is armed to the teeth, starts high, does his trick and lights the scat power to get out of there. A C-130 pilot does his act low and slow and wishes he had 'just a little' scat power. If things don't go right for the F-4 pilot, he can wait for his seat to kick him out into the blue. The C-130 guys have to unbuckle, run about 50 feet to the cargo ramp, and then decide if a parachute landing fall is better than riding the beast into the ground."

Both strategic and tactical transports in Vietnam ended up carrying everything from tanks and ammunition to elephants and refugee Vietnamese babies.

On January 1, 1966, MATS was redesignated the Military Airlift Command, with its strategic airlift and other missions remaining mostly the same. However, the Air Communications Service Mission had been separated from MATS in 1961, and MAC's Air Rescue mission was now expanded to include coordinating search and rescue missions within the United States as well as supporting NASA's space explorations. Air Rescue was renamed the Aerospace Rescue and Recovery Service. The Air Photographic and Charting Service which had begun under MATS in 1951 became the Aerospace Audiovisual Service, providing motion picture, television, and still photographic coverage for all Air Force activities.

In December, 1974, MAC assumed responsibility for tactical as well as strategic airlift. By consolidating strategic and tactical airlift under one command, the efficiency and flexibility of total airlift was greatly increased.

In 1977, MAC became a specified command, still under USAF but reporting directly to the President through the Secretary of Defense and Joint Chiefs of Staff during wartime and periods of crisis. Special Operations Forces was added to MAC in 1983.

MAC's many, versatile missions were tested in October, 1983, during the Grenada operation called "Urgent Fury." Grenada involved almost every MAC wing in the United States, from special operations, aeromedical evacuation, aerospace rescue and recovery, to weather and audiovisual services. MAC planes airlifted

25

Army troops march off a MAC C-141B Starlifter after returning to Pope AFB in North Carolina from Grenada.

troops, military equipment, and helicopters to the Point Salines airfield on Grenada, and MAC planes carried the wounded and assault forces home.

From its original mission of operating as little more than a militarized commercial airline, MAC, today, has become the "backbone of deterrence" for U.S. fighting forces. Wherever America's vital interests are threatened, MAC must be ready to deploy, resupply, and redeploy U.S. combat forces and their support equipment anywhere in the world, and do it in a matter of days.

The people of MAC, however, are not just trained to operate in wartime and under battle conditions. MAC also is the largest peacetime cargo airline in the world, its customers the Department of Defense and its components.

In addition, MAC also operates the world's largest and most far-reaching humanitarian airlift, throwing a lifeline to peoples threatened by natural or man-made disasters. As we shall see in the next chapter, perhaps no other military organization has touched the lives of so many people of so many lands.

With the delivery in 1969 of the C-5 Galaxy, the world's largest aircraft, MAC achieved a revolution in airlift. Here an F-5 is being loaded aboard a C-5.

CHAPTER 2

Lifeline in the Sky

It was an early Friday afternoon in October, 1980, when the ground began to shake beneath the city of El Asnam in Algeria. Within minutes, the earth heaved and office and apartment buildings, hotels, schools, and mosques collapsed like houses of cards. Frightened residents rushing out into the street were crushed beneath the bricks and stones. One survivor said, "Everything happened so quickly. The dogs did not have time to bark." When the last tremors had ended, 80 percent of the city was destroyed and thousands were dead or injured, still trapped beneath the rubble.

By Sunday morning the first MAC C-141 had arrived at Algiers carrying relief supplies and a 38-man Disaster Assistance Survey Team. By the time the airlift was finished, eight MAC C-141s, three C-130 Hercules, and one C-5 Galaxy aircraft and their crews had flown 375 tons of tents, blankets, and other supplies to the devastated city.

Earlier in the year, in August, Hurricane Allen, the second most powerful Atlantic hurricane in history, ripped through the Caribbean, leaving shattering death and destruction in its wake. Allen's winds were clocked at 185 miles per hour, demolishing thousands of homes and killing more than 100 people. In just six hours, the hurricane reshaped the coastline of Jamaica. Two beachfront hotels disappeared with a single slap from 30-foot waves.

Immediately MAC dispatched Disaster Assistance Support teams, helicopters, and supplies to the stricken islands of Jamaica, Haiti, Barbados, Dominica, and St. Lucia.

One of the longest continuing humanitarian relief missions performed by MAC was the African drought relief in 1973 and 1974. Rainfall dropped well below normal in the sub-Sahara region, turning the land into a dust bowl, incapable of supporting grain or livestock. Wells dried up and the people in the 7,000-mile region faced death by dehydration, starvation, or disease. Since there was no railroad and only dirt roads into the crowded refugee camps in the interior, airlift was essential to bring in the food necessary to keep millions of people alive.

Joining with other countries, the United States answered the urgent call of the United Nations international African relief effort. The Military Airlift Command airlifted tons of food, primarily wheat, rice, sorghum, and powdered milk for the children, along with goats, sheep, and water buffalo to the drought-stricken African countries.

The airlift operation was complicated by scorching 115-degree temperatures over the Sahara Desert and red sandstorms often climbing to 10,000 turbulent feet, making visibility difficult. There

was no difficulty, however, in off-loading the food! Starving men and women scrambled aboard the planes and the last heavy sack of grain would be out of the aircraft in 25 minutes. Then the floor of the aircraft was carefully swept so that not a grain of food would go to waste.

Ten years later an even worse drought and famine devastated Africa again. In Ethiopia alone 300,000 died. Once again food and supplies were sent to Africa from countries and relief agencies all over the world. MAC's C-141s airlifted in blankets, measles vaccine, skim milk, water tanks, and other vital relief supplies to the human wave of refugees fleeing to Sudan from Chad and Ethiopia.

Through the years the Military Airlift Command has thrown an airlift lifeline countless number of times to countries ravaged by earthquake, flood, hurricane, famine, and other natural disasters. Usually the request for disaster relief comes first from the U.S. Ambassador in the stricken country to the State Department, who passes the request along to the Secretary of Defense. If airlift is deemed necessary, the Joint Chiefs of Staff then tasks the Military Airlift Command.

Natural disasters, of course, do not happen only in foreign countries. Within the United States, MAC delivered relief personnel, emergency equipment, and supplies after a massive earthquake split the earth and sent buildings tumbling in Alaska. Food and supplies were airdropped to 50,000 Navajo Indians stranded by snow in the mountains of northwest Arizona.

MAC airlifted almost four hundred tons of sandbags to Min-

A MAC C-130 departing after bringing grain to Africa during a period of famine.

Snow removal equipment is off-loaded from a MAC C-130 during the blizzard of 1977 in Buffalo, New York.

nesota to fight off rampaging floods there, and in Arizona, fire-fighters were airlifted in from neighboring states and chemical fire retardants dropped to control forest fires raging throughout that state.

Not all disasters are natural. Many are man-made. Political upheavals around the world have brought about civil wars bringing death and destruction in their wake. Military airlift can sometimes stop these small wars from turning into worldwide conflagrations, provide military material to allies, and deliver food and medicine to victims of war.

Two days after the Belgian Congo (later renamed Zaïre)

32

gained its independence in 1960, fighting broke out between out-law rebel forces. Europeans and Americans trapped within the country, as well as Congolese citizens, were savagely killed or taken hostage and threatened with execution. President Lumumba requested military aid from the United Nations to restore order. Less than 48 hours later, MAC (at that time, the Military Air Transport Service) and U.S. Air Forces in Europe mounted what would become by the end of 1960, the largest American military airlift since the Berlin blockade.

It was an airlift made more difficult by aircrews operating in the harsh, unfamiliar environment of equatorial Africa. Navigation maps were unreliable, showing mountains where there were no mountains or marked in the wrong places. Celestial navigation was hampered by intense desert sandstorms that hid the stars. Radio beams were of low frequency or nonexistent, and multilingual air traffic controllers had to be found in a region where English was seldom spoken.

Language was also a confusing barrier in transporting United Nations soldiers from sixteen different countries to the Belgian Congo. Many of the troops were traveling in airplanes for the first time in their lives. By the time the airlift ended in January, 1964, MATS had flown some 2,000 missions, moving 46,000 United Nations troops and more than 10,000 tons of cargo. Despite the difficulties, the foreign troops and field equipment were airlifted successfully without a single serious accident or incident.

Other military airlift missions to political hot spots around the world followed in the 1960s and '70s. MAC airlifted Air Force and Army units and supplies during the Cuban and USS *Pueblo* crises, evacuated to safety U.S. and other foreign nationals from the Dominican Republic and Pakistan, brought in a contingent of peacekeeping forces to Zaïre after an invasion by rebel troops from Angola. More and more, airlift began to take a leading role in projecting America's military force abroad.

The most vital of these military airlift missions, backing up U.S. national interests, was the Israeli airlift of 1973, during the Yom Kippur War. A fierce conflict had erupted between Israel and Egypt and Syria. Russia was supplying Egypt and Syria with arms, and Israel was running dangerously low on tanks, rockets, and ammunition.

Within nine hours of the U.S. decision to assist Israel, the first C-5 aircraft was loaded and airborne. Within 33 days, MAC C-141s and C-5s flew 566 missions over 6,450 nautical miles, making only a single stop en route for refueling. More than 22,000 tons of critical war material was airlifted to Israel, effectively turning the tide of battle. By comparison, over a 40-day period, the Soviet Union airlift provided Egypt with 15,000 tons of war material in 935 missions while flying a much shorter route of only 1,700 nautical miles. And MAC successfully completed the Israel airlift while routinely performing its other many daily regular missions.

Airlift missions in response to a sudden military or humanitarian crisis are called Special Assignment Airlift Missions (SAAMs). SAAMs are also operated when a mission requires a special pickup or delivery at points outside the established MAC routes, such as airlifting military troops and cargo in support of military exercises, or when other airlift or transportation means are inadequate.

Although the demands of each SAAM are different, SAAMs, in response to military or humanitarian crises, have one element in common—the need for a swift, efficient response to the emergency without jeopardizing MAC's business-as-usual, regular airlift missions.

On a typical day, more than 170 MAC aircraft are away from home station carrying passengers and a cargo to as many as twenty countries, with more than 800 arrivals and departures in a 24-hour period. MAC's customers are the Air Force, Army, Navy, Marines, and other Defense Department agencies and commands. These missions can include everything: resupplying units stationed at the Antarctic for Operation Deep Freeze, airlifting cargo to a remote Navy base, transporting military men to new assignments overseas, flying around the globe twice a week carrying embassy pas-

Former American prisoners of war in North Vietnam are airlifted aboard a MAC C-141, headed for Clark Air Base, and then home, in 1973.

sengers and mail, or making a daily resupply run to the middle and western Pacific, among many, many other routine missions.

In one year's period, MAC carried more than 2,100,000 passengers and 489,000 tons of cargo through its aerial ports—its passenger and cargo terminals. This type of airlift is known as "channel," which is regularly scheduled service, as opposed to Special Assignment Airlift Missions (SAAMs). It is no wonder that MAC is called the "biggest, busiest, most far-ranging aerial cargo carrier in the world!"

The MAC Command Center, located in the big, red brick headquarters building at Scott AFB, Illinois, is the hub for MAC's worldwide airlift operations. As part of his morning briefing at the Center, the commander in chief of MAC is informed of the status of the more than 1,000 aircraft belonging to MAC, as well as all airlift missions currently in progress.

A network of subordinate command centers at MAC bases within the United States and overseas directs and operates airlift missions within their own areas. These centers are connected with MAC Command Center through command-wide computers, and headquarters at Scott AFB is constantly being provided with updated information on every airlift mission around the globe.

Since MAC's aircraft may have to fly into military bases and civilian airports anywhere in the world, ALCE (Airlift Control Element) teams are sent out first from the nearest U.S. or overseas subordinate command center, to act as mobile command centers at those locations.

On a recent NATO military exercise when MAC planes with assault troops flew into a small Norwegian air base, an ALCE team was waiting at the base. The ALCE cadre helped unload the planes, found quarters for the aircrews, repaired aircraft, organized flight plans, prepared crew briefings, and took care of a hundred other details involved in an airlift mission into a foreign air base. The ALCE even persuaded the Norwegians to cut down trees within 75 feet of the taxiway to get necessary wing clearance for the C-141s that were due in the next day!

When orders come into MAC Command Center for a Special Assignment Airlift Mission in response to an emergency situation, a Crisis Action System is activated and a Crisis Action Team (CAT) is formed. The team, which is on standby at all times, reports within one hour to the Command Center. Each member of the Crisis Action Team is a specialist, in maintenance, transportation, communications, personnel, or operations, among other fields.

MAC is divided into three subordinate numbered Air Forces: the 21st, with headquarters at McGuire AFB, New Jersey, covering the eastern hemisphere; the 22d, headquartered at Travis AFB, California, covering the western hemisphere; the 23d, located at Scott AFB with a worldwide command over rescue, aeromedical evacuation, and special operations, among other tasks.

The Crisis Action Team at headquarters must decide which divisions and wings, which bases and personnel, and which reserve forces, if any, will be involved in the emergency airlift mission. Each wing tasked has its own Crisis Action Team.

The headquarters team must make a great many other decisions. Which aircraft routes are to be flown? What en route support requirements, such as refueling for aircraft, will be needed? Are adequate airfields available to MAC aircraft in the foreign country, or will airdrops be necessary to deliver the supplies? And if airdrops are necessary, what special problems will be faced in weather and terrain? Communication systems, using satellite radio communica-

American hostages in Iran return home on a MAC C-137 in January of 1981 after nearly a year and a half of captivity.

tions, are set up or restricted, as necessary. Commercial airlines may be called in to free military aircraft to participate in the crisis airlift.

The Crisis Action Team at wing level must act quickly in assigning crews and aircraft with the needs of the emergency, for rapid and efficient delivery of assistance to the victims may mean life or death to many. The CAT teams operate around the clock until the last plane returns to the base. However, the crisis itself as it develops is often unpredictable, requiring sudden, unexpected changes in planning and operation.

The tragedy at Guyana in 1978 was one such unpredictable

38

airlift mission. At 8:30 P.M., on Saturday, November 18, MAC was alerted by the National Military Command Center in the Pentagon that a U.S. congressman and several American citizens had been murdered at Jonestown, the colony of an obscure American religious cult. A Crisis Action Team was immediately assembled at MAC headquarters. Six hours later a C-141 was dispatched from the 437th Military Airlift Wing at South Carolina, carrying, among others, an aeromedical evacuation team and a MAC Combat Control Team to provide security.

After arriving at Guyana, the nightmare of the massive suicides that had happened at the Jonestown colony was revealed. A Joint Air Force and Army Task Force was set up. The Crisis Action Team at MAC headquarters had to enlarge its mission into a full-scale airlift.

It was immediately apparent that heavy-lift helicopters were needed to shuttle out the bodies of the Jonestown victims. Three HH-53 helicopters were dispatched from the 55th Aerospace Rescue and Recovery Wing at Eglin AFB, Florida. The helicopters required aerial refueling by HC-130 aircraft several times before they reached Guyana. Nine C-141 flights airlifted the bodies from Georgetown, the capital city, to Dover Air Force Base in Delaware. Also needed were consular officials, medical and graves registration teams, communication gear and specialists, support troops and supplies.

Before the MAC Crisis Action Team finally stood down 166 hours later, thousands of people, both Air Force and Army active and reserve personnel, scores of air bases and almost seventy transport missions were involved in the Guyana airlift.

No Crisis Action Team, of course, works alone. Supporting them in their mission are the 93,000 active-duty military and civilian personnel of the Military Airlift Command as well as crewmembers, ground personnel, and aircraft of the Air Force Reserves and Air National Guard. MAC personnel may be found at 340 locations in 26 countries, with 14 MAC bases in the United States and two MAC-controlled facilities in Europe, one in Germany and one in the Azores.

No group within MAC, though, provides more specialized and vital assistance in a natural or man-made disaster than the highly trained personnel described in the following chapter—the airmen and women of MAC's Aerospace Rescue and Recovery Squadrons.

CHAPTER 3

"That Others May Live"

Deep within Viet Cong territory, an American pilot is downed by enemy fire during the Vietnam War. The enemy soldiers close in, hoping to use the survivor as "flak bait" to shoot down helicopters coming to rescue the man.

Responding to the pilot's Mayday signal are A-1 Skyraiders and a HH-3E Jolly Green Giant helicopter of the Aerospace Rescue and Recovery Squadron. After the Skyraiders have laid down a smoke screen, Airman Duane Hackney, a pararescueman aboard the rescue helicopter, volunteers to be lowered into the jungle to search for the downed pilot. After two sorties, Airman Hackney locates the pilot, who is hoisted into the helicopter.

As the rescue crew departs the area, intense and accurate 37-mm flak tears into the helicopter amidship, causing damage and a raging fire aboard the craft. With disregard for his own safety, Air-

41

man Hackney fits his own parachute to the rescued man. Locating another parachute for himself, Airman Hackney manages to slip his arms through the harness when a second 37-mm round strikes the crippled aircraft, blowing Hackney through the open cargo door. Though stunned, the pararescueman manages to deploy his unbuckled parachute and make a successful landing. He is later recovered by a companion helicopter.

For his work that day, Duane Hackney received the Air Force Cross. But Hackney was not alone in his heroism. Pararescuemen of the Aerospace Rescue and Recovery Service—called PJs for parachute jumping—won more decorations than any other group of men in the Air Force serving in Vietnam.

An organized military effort to rescue downed airmen, however, did not start with the Vietnam War. Germany, using the Heinkel-59 float plane, first pioneered air-sea rescue of its Luftwaffe aircrews downed in the English Channel during World War II. The English and Americans soon followed with their own air-sea rescue teams. Modified American B-17s parachuted plywood lifeboats, stocked with supplies, to aircrew survivors.

In August, 1943, when a C-46 crashed over an uncharted jungle near the China-Burma border, the only means of getting help to the survivors was by paradrop. A lieutenant colonel and two medical corpsmen volunteered for the assignment. For a month, these men, the first PJs—although the Air Rescue Service had not yet been founded—cared for the injured until the party could be brought to safety. The success of this first parachute rescue team proved that a highly trained rescue force could save survivors when a plane was downed.

Demonstration of HH-53 helicopter hoist used in rescue operations.

After the war, the question arose as to which service should be responsible for rescue operations. Rescue at sea had always been the traditional responsibility of the United States Coast Guard. The Army Air Forces, however, wanted to expand their own air rescue capabilities. In 1945, it was decided that the Air Transport Command, forerunner of the Military Airlift Command, be given the responsibility for air search and rescue over land and ATC's overseas air routes.

When North Korea invaded South Korea, the new Air Rescue Service moved swiftly into action, first with its Sikorsky H-5 helicopters and, later, the larger and faster H-19. The H-5s and Army rescue helicopters not only moved wounded soldiers from the battlefront to the Mobile Army Surgical Hospitals (MASH), but assisted in evacuating troops trapped behind enemy lines. Helicopters performed most of the rescue work in Korea, although there were also fixed-wing aircraft involved. The lumbering, slow-moving Grumman SA-16A Albatross, called Dumbo, could operate from both land and water.

By the end of the Korean War, rescue crewmen of the 3rd Air Rescue Group were credited with airlifting 9,680 military personnel to safety, as well as 996 combat saves. The Air Rescue Service had become a vital part of the United States Air Force, along with its new special breed of man—the pararescueman.

It was the Vietnam War, though, which developed the rescue of downed pilots on sea and land to a fine art. As more and better-trained pararescuemen moved into Vietnam, the helicopters used in rescue missions were also improved.

One of the improvements that revolutionized search and rescue operations was the aerial refueling of helicopters. The first in-flight transfer of fuel between an HC-130P and an HH-3E happened in

1966. In-flight refueling extended the range of the helicopters, allowing them to continue circling, and cutting the time it took to reach airmen down in North Vietnam and Laos.

However, despite the advantage of aerial refueling, early rescue helicopters, such as the HH-43B/F Huskies and HH-3E Jolly Green Giants, lacked sufficient armor to survive the intense anti-aircraft fire in North Vietnam, and their engines were not powerful enough to maintain a hover over the jungles of the higher mountains. In addition, armed with only a 7.62-mm machine gun, the choppers did not have enough firepower to "shoot their way out" of dangerous situations.

Then, in 1966, the Sikorsky HH-53B, twice the size of the HH-3E, was developed. Dubbed the Super Jolly Green Giant, the

An HH-53 Super Jolly Green Giant being refueled by an HC-130. With air-to-air refueling capability, the range of the HH-53 is limited only by the crew's endurance.

HH-53 could carry a crew of six, including two pararescuemen, and 40 fully-equipped soldiers, if necessary. Its two GE-T64-3 turboshaft engines, speed of 195 mph, titanium armor, and three 7.62-mm Gatling type miniguns, as well as its air refueling capability, made the HH-53B the largest, fastest, and most powerful helicopter in the Air Force.

The rescue of a downed airman in a hostile environment, however, requires more than courageous pararescuemen and powerful helicopters. The downed airman must first be located, his position verified, and his physical condition determined. The location of the nearest enemy in the area must also be discovered. If rescue appears feasible, airborne command and control of the mission must be coordinated with ground controllers at rescue centers to determine the type of rescue mission that will be undertaken.

A typical search and rescue might be as follows. After being hit by ground fire, a pilot ejects from his fighter plane over North Vietnam. Usually he lands in a tree above the jungle floor, suspended by his parachute. After cutting himself loose with his survival knife, he uses his URC-11 survival radio in his seat pack to contact an airborne rescue command post. One or several A-1 Skyraiders respond to his Mayday within half an hour.

The Skyraiders radio the downed pilot's exact position and then fly to another area, circling several miles away so as not to reveal the pilot's location to enemy troops lurking nearby. If the location of the enemy is uncertain, the Skyraiders make low, level passes over the area for several hours, a tactic called "trolling for fire" until the enemy forces are located.

When the rescue helicopter arrives, the downed pilot fires off a small flare and the chopper moves directly overhead while a PJ lowers the jungle penetrator. The penetrator has spring-loaded

HH-3 helicopter lowering a forest penetrator to hoist a pararescueman and
a downed airman. Penetrators were used in the jungles of Vietnam.

A double hoist from water to air by pararescuemen.

arms that part the jungle foliage as it is lowered. The survivor straps himself to the penetrator and releases a set of spring-loaded arms at the other end for protection as he is hauled up through the branches of the trees. If the pilot is injured, a pararescueman will lower himself on the penetrator and assist the pilot up to the helicopter and tend his wounds.

If a pilot is shot down over water, the Albatross, a fixed-wing SA-16A first used in Korea, often becomes the rescue vehicle. The amphibious plane can make a water landing to pick up survivors, if the sea permits, or lower a hoist to the downed pilot if the sea is too rough. If necessary, the pararescueman can parachute into the water to lend assistance.

Whatever the procedure used, by 1966 a downed aircrew in Vietnam had a one-in-three chance of rescue. By the time of America's withdrawal from Vietnam, the Aerospace Rescue and Recovery Service, formerly Air Rescue Service, had saved 3,883 lives.

The Aerospace was added to Air Rescue's title when space exploration added a new duty to their mission, that of retrieving nose cones, space capsules, and astronauts in support of NASA manned-space missions. When the decision was made to terminate the Gemini 8 space flight in 1966, making an emergency splashdown about 500 miles east of Okinawa, a rescue aircraft crew arrived in time to see the spacecraft hit the water. Three PJs parachuted into the ocean and had flotation equipment attached to the spacecraft within 20 minutes. They stayed with the astronauts until a Navy destroyer arrived three hours later.

In addition to wartime combat rescue and rescue coverage for manned space flights, ARRS provides flight crews for another Military Airlift Command service: weather reconnaissance. But

ARRS's best-known peacetime mission is its responsibility for military and civilian search and rescue due to natural or man-made disasters.

Such rescues can range from pararescuemen descending from helicopters to snatch four frostbitten and injured mountain climbers trapped on Mount McKinley to saving 74 lives when fire broke out aboard the cruise ship *Prinsendam* in the Gulf of Alaska. At 6:00 A.M., passengers were ordered to abandon the sinking ship. By 9:30 A.M., helicopters from rescue units in Alaska were hoist-lifting survivors from the lifeboats and ferrying them to rescue vessels.

Within 75 minutes of the volcanic explosion at Mount St. Helens, Washington, Air Force Reserve rescue helicopters were on their way to remove survivors from the mountain, with a total of 101 lives saved.

A small plane crashes in the wilderness. Within two hours, pararescuemen reach the crash site and give medical aid to the two survivors.

A sailor is badly burned on a Russian ship in the Atlantic, 700 miles from the nearest land. Two PJs, stationed in the Azores, are flown to the Russian ship. They parachute near the ship, are picked up, and provide medical treatment for the sailor until the ship reaches port days later.

No matter how small or large, all federal land and sea rescue operations within the continental United States are coordinated and controlled through the Air Force Rescue Coordination Center located at Scott Air Force Base, Illinois. The Center is manned by personnel trained in search and rescue operations and is equipped with telephone, teletype, air-to-surface radio, and computer capability.

When an emergency is reported to the Center, the incident is verified and Center workers place a numbered arrow on a large map on the wall in front of the communications consoles. A notation is made on a blackboard at the side. A dozen or so rescue efforts may be in progress at one time at the Center, which operates around the clock.

The request for assistance is immediately passed to the appropriate rescue agency, whether it is local, state, or federal. Such rescue agencies include everything from the local Civil Air Patrol to a squadron of the ARRS. The emergency can be anything from lost skiers or hikers to the transporting of a human eye by helicopter from one hospital to another, to a flooded river threatening a town, or an overdue light plane.

Speed is essential in any rescue attempt, but in the case of plane crashes, studies show that the survival rate is better than 50 percent if rescue can be accomplished within eight hours.

To that end, international Search and Rescue Satellites now monitor emergency locator transmitters worldwide. The satellites of both the United States and Russia "listen" for distress signals from aircraft and ships, relaying such signals immediately to a network of ground terminals. The information is relayed to the United States Mission Control Center, which shares space with the Air Force Rescue Coordination Center. During its first year of operation, the Search and Recovery Satellite Aided Tracking program (SARSAT) contributed to saving more than 90 lives throughout the world.

In peace and in war, the Aerospace Rescue and Recovery Service has saved more than 20,000 lives. Its pararescuemen are among the most highly trained, dedicated professionals in the armed forces. How does one become a pararescueman? Well, it isn't easy!

51

Pararescue men line up for their turn to use their parachutes.

All pararescuemen are volunteers and all must successfully pass a grueling test before entering pararescue training. Of every 20 volunteers, usually only six or seven pass the test. The training begins at Lackland AFB, Texas, with a rigorous eight weeks of physical conditioning and discipline. The student's day begins at 5:00 A.M., and can include a 1500-meter swim, a two-mile run, and two hours of calisthenics, or a seven-mile run with pushups, and situps, or a 4000-meter swim plus pool harassment, underwater

A pararescueman, already trained in the air, takes to the water with scuba equipment.

work, and more calisthenics. A typical class of 40 trainees may graduate only 12 students.

Then follows three weeks of jump school at the U.S. Army Airborne School at Fort Benning, Georgia. This is followed by what many consider the most difficult part of the training: five weeks at U.S. Army Special Forces Scuba School at Key West, Florida. After the scuba training, there are three weeks at the U.S. Basic Survival School at Fairchild AFB, Washington, learning how to survive in the desert, jungle, swamps, and the arctic. Finally, there are 18 weeks at the Pararescue Recovery Specialist Course at Kirtland AFB, New Mexico, putting all the newly learned skills together, including qualifying as a medical technician.

At his graduation, the pararescueman is ready to take his place among that select group of men of the Aerospace Rescue and Recovery Service who have risked their lives in peace and war to save life and give aid to the injured, fulfilling the proud motto of AARS: "These things I do—that others may live."

CHAPTER 4

The Hurricane Hunters

The Air Weather Service, a technical service of the Military Airlift Command, studies weather and environment all around the earth and in outer space. But for the pilots and crews of MAC's Weather Reconnaissance Squadrons, whose mission is to fly into some of the most terrifying and destructive natural forces known, the focus can be much more limited. The target of the men and women on these weather reconnaissance flights is the eye—the calm center—of a hurricane or typhoon.

The 53rd Weather Reconnaissance Squadron (WRS), along with an Air Force Reserve WRS, both stationed at Keesler AFB, Mississippi, are USAF "Hurricane Hunters" or "Storm Trackers" in the Atlantic, Caribbean, Gulf of Mexico, and eastern Pacific. The 54th WRS, flying out of Guam, performs the same task as "Typhoon Chasers" in the western Pacific Ocean. A combined Air

Weather Service and Naval Oceanography Command provides a Joint Typhoon Warning Center in the Pacific.

Let's go on one "hurricane hunting" flight involving Camille, the greatest recorded storm to ever hit a heavily populated area of the Western Hemisphere. The story is summarized as told by a copilot of the 53rd Weather Reconnaissance Squadron.

On August 14, 1969, the National Hurricane Center in Miami reported that the storm Camille in the Caribbean had become a hurricane and had changed its progress. The crew of the WC-130, a weather-modified version of the cargo-carrying C-130 Hercules, was told to get "eye" penetration of the hurricane and exact position fix at a set time. As the copilot of the WC-130 put it, "With Jamaica behind us, the granite faces of thunderstorms in the first line of feeder-band clouds tightened our already tense stomachs. I

Present-day "Hurricane Hunters" use the WC-130, a modified version of the Hercules. It can penetrate hurricanes at 10,000 feet to collect data from the eye of the storm.

Navigator keeps track of the aircraft's position, using radar to locate the storm and guide pilot into its eye.

was in the left seat, flying the airplane through an area where common sense warned that we should not go."

Using compass headings selected from radar returns, the WC-130 was put on an irregular course through thunderstorms, violent turbulence, and torrential rain. Finally, they flew into an area clear enough to provide a view of the rough sea 10,000 feet below. The navigator came through on the intercom: "I've got a clear radar return ahead. Fifty miles. The classic shape of the eye!"

All around the plane were steel-blue sky walls of dense thunderstorms and strong winds. The aerial reconnaissance weather officer found the velocity of surface winds was 80 to 90 miles per hour,

Dropsonde operator loads a drop-chamber capsule with a radiosonde transmitter aboard a WC-130 aircraft.

well in excess of hurricane wind speeds beyond 73 miles per hour. The navigator sought a soft spot in the "wall" surrounding the eye, and found a narrow opening between towering thunderstorms. Their radar was "drowned" by the water; they hit deafening noise and more turbulence. Then the navigator called, "We're inside. I've got the radar back; the eye is real tight."

In the warm center of the eye, the point of lowest barometric pressure, the dropsonde operator dropped his instrument package, a two-foot by six-inch, parachute-guided, nonrecoverable tubular sensing device. It fell 5,000 feet per minute, sending radio signals that gave barometric pressure, temperature, and humidity from the aircraft to the ocean surface. All of the data was transmitted immediately to the hurricane center.

The WC-130 circled elliptically within the eye, in case the dropsonde unit failed, while the Weather Officer read the way the waves were breaking to obtain surface wind directions. Such data is especially critical as a storm tracks toward populated land areas.

With a good readout from the dropsonde, the WC-130 was "crabbing" its way back through the wall cloud when the navigator called for an immediate hard left turn to avoid a possible tornado. Suddenly, radar and shortwave radios went out. Without them, they could not make a second penetration of the eye. Using Very High Frequency radio, limited to a few line-of-sight stations in Cuba and Mexico, they got clearance to fly to Florida. Flying blind in high cirrus clouds, their Hercules was hit by lightning and their dome protector for the radar burned. They managed to land at McCoy Air Force Base, Florida.

With radar repaired, the original crew took their Hercules up once more, hunting the eye of Camille. Winds over 200 miles per hour now could be determined by the deeply wind-furrowed sea below. At the plane's altitude, the winds were 190 miles per hour, a higher velocity than they had ever before recorded. Lurching into the eye, they "kicked out" the dropsonde unit and orbited, circling for an accurate readout.

When oil pressure on their number three engine all at once dropped toward zero, they got out of the eye and sought a previously seen 50-mile clear area. Their number four engine light indicated a faulty generator. They shut number four down, finally landing safely in Houston, Texas.

The MAC Weather Reconnaissance Squadron was not alone in its hurricane hunting that day. A Navy DC-121 also flew into the eye of that storm. But this is not just an adventure story. The data gathered by the aerial weather squadrons showed that Camille was within a day of striking Louisiana, Alabama, and Florida. Because of the convincing measurements of the Hurricane Hunters, over 200,000 people left Camille's path and found safety. Some stayed; 255 were confirmed killed and 68 missing.

The first low-level penetration of a hurricane by Air Weather Service aircraft happened in October, 1947; the first night penetration occurred in August, 1955. Both times the aircraft was a WB-29.

In the twenty years before aerial weather reconnaissance, an estimated 6,200 persons lost their lives in the United States from killer hurricanes. On Labor Day, 1935, a hurricane over the Florida Keys killed between 400 and 500 people. In September, 1960, however, a similar hurricane over the Florida Keys killed only three, although it was one of the most destructive hurricanes of all times. Radar and aircraft warnings of the approaching hurricane were given in time for the people in the path of the storm to be evacuated to safety.

In the last two decades, mostly because of the men and women who work in the Air Weather Service and Aerospace Rescue and Recovery, fatalities from hurricanes in the United States have been reduced to an estimated 1,400.

As important as the Air Weather Service is to the civilian population, though, AWS's primary mission is to operate a worldwide network of weather facilities, providing weather support to the Air Force, Army, and the Department of Defense in peacetime and war.

60

Weather has always been a deciding factor in battles. A snowstorm and a half-frozen Delaware River formed obstacles but aided vital surprise when General Washington captured the Hessian forces at Trenton in 1776. Torrential rains forced General Burnside to abort his infamous "Mud March" at the important Civil War Battle of Fredericksburg in 1863. But weather observation did not become an official part of the military until 1870 when President Ulysses S. Grant added a weather section to the Army Signal Service.

The first full-scale employment of United States military weathermen came during World War I, when one historian said that "the battles were almost as much against the weather and the mud as against the Germans." The science of meteorology during the First World War, however, was still very new and the weather information was used chiefly by the artillery, the fledgling air corps, and in planning poison gas attacks.

The Army Air Forces Weather Service was established in 1937 to support the emerging military aviation operations. By 1945, some 19,000 military weather professionals were working at more than 900 locations throughout the world.

In World War II, weather forecasters on both sides, American and German, were determining the time and location of air raids to avoid the worst weather or to take advantage of good weather conditions. The date of D-Day for the Allied attack on the Germans in Normandy, France, the greatest military operation ever mounted, was decided by weather consideration.

In the Pacific theater, the Japanese took advantage of an extensive storm zone in the Pacific to conceal the approach of their aircraft carriers to Pearl Harbor in December, 1941, and basically, the ending of World War II with Japan came with the weather-determined timing of the dropping of the atomic bomb.

After World War II, in 1946, the Army Air Forces Weather Service was renamed Air Weather Service and assigned to the Air Transport Command. When the ATC became MATS and then MAC, the Air Weather Service became responsible for providing weather support to the Air Force, the Army, and joint and combined operations in war and peace.

During the Korean War, weather reconnaissance missions were flown daily over North Korea with specially instrumented aircraft manned by Air Weather Service trained personnel. By the end of the war, the 56th WRS was the only Air Force unit to have had an aircraft over enemy-held territory every day since the war began.

By the time of the Vietnam War, Air Weather Service had established an Air Force weather satellite system. This new weather-observing tool, the meteorological satellite, proved invaluable to the weathermen in Southeast Asia. Satellite pictures became the primary source of determining the cloud conditions in a target area, such as the Son Tay POW raid and the reopening of Khe Sanh.

Weather reconnaissance aircraft flew daily missions over North Vietnam, gathering weather data, as they had in Korea. On the ground, Air Force weathermen, providing weather support, were often deployed forward with Army combat battalions on "search and destroy" operations. There were even Special Warfare weather teams that worked clandestinely in Laos under dangerous condi-

Swirling cloud banks in weather satellite photograph show low pressure area off Virginia coast and heavy thunderstorms over south central United States. Weather satellites help predict possible tornados and hurricanes.

tions, to establish and maintain weather observations essential to combat air operations.

In Vietnam, weathermen did not only forecast rain. They made rain—on request. From 1967 to 1972, the Air Force WC-130s and RF-4Cs dropped silver iodide, seeding clouds to cause rain. By creating mud on the ground, they slowed the flow of enemy supplies. This humane "weapon" saved lives at relatively low cost.

Until 1975, the Air Weather Service alone was responsible for flying weather reconnaissance aircraft. At that time, responsibility for flying the weather-modified WC-130s was transferred to the Aerospace Rescue and Recovery Service. The Air Weather Service still provides two crewmembers: the aerial reconnaissance weather officer and the dropsonde operator.

Formerly, many weather reconnaissance squadrons were stationed around the world, but with the successful deployment of weather satellites, the need for aircraft-supplied weather data has decreased. However, after the satellite detection, pinpointing hurricanes still requires the expertise of aerial reconnaissance units.

Today, all types of meteorological equipment are used by the Air Weather Service, from simple wind gauges to weather radars and solar telescopes. But weather forecasting still requires masses of accurate and timely raw data, collected on a worldwide basis from Air Force and Army installations all over the world.

So much data can only be handled efficiently by computer technology. The Air Force Global Weather Central (AFGWC), part of the Air Weather Service, is located at Offutt Air Force Base, Nebraska, and has the largest military meteorological computer facility in the world. More than 140,000 weather reports per day are received, processed, and analyzed by computers, including information gathered by the Defense Meteorological Satellite Program. Forecasters are on duty 24 hours a day and can issue weather forecasts for anywhere on the globe for many types of military operations.

The Automated Weather Network, a high-speed digital communications system, connects AFGWC with U.S. military users in fifteen countries.

With the new weapons and vehicles of the space age, such as ballistic missiles and space shuttle flights, AWS also requires a knowledge of weather conditions in and beyond the earth's atmosphere.

To support weapons and vehicles in outer space, the AWS Space Environmental Support System has been established. The system is a worldwide network of solar observing sites. The sites are equipped with solar, optical and/or radio telescopes, weather satellites, storm detection radar, rocketsondes (small weather

Preparing a rocketsonde for firing. The rocketsonde provides weather sensors from 12 to 50 miles in altitude.

rockets), balloon rawinsondes (weather instruments carried aloft by balloons), and various other solar sensors to monitor the sun and near-earth space environment. With this data fed to Air Force Global Weather Central, predictions can be made of solar-geophysical events that will affect military systems operating in deep-space environment.

Air Weather Service responsibility does not end with forecasts to the Air Force and Army. The mass of data accumulated daily at AFGWC is retained as a working, historical file at the USAF Environmental Technical Applications Center, Scott AFB, and an operating location at Asheville, North Carolina.

At these two locations, AWS climatologists study climate, or the weather conditions over a period of years in various parts of the world, determined by temperature and meteorological changes. This information on climate and the environment is valuable to the Department of Defense in its planning for military operations and in the acquisition of new weapons systems.

Through working with national and international committees, the AWS also helps provide weather support for the fighting forces of the free world. These committees exchange weather data and coordinate plans for peacetime and wartime weather support between nations.

The AWS is constantly developing new techniques and instruments to study weather and climate, taking advantage of new computer and communications technology. Soon it will have the ability to provide a three-dimensional picture of the atmosphere, and a more highly developed Doppler weather radar will provide accurate, timely warnings of severe and hazardous weather. Battlefield weather observation and forecast systems will be able to collect

66

weather information from hostile areas and transmit them immediately to commanders in the field.

The war the AWS fights is the oldest known, a fight that has continued down through human history. The fight takes place in peace and war, and the battles never cease. All human "victories" are temporary when fighting weather. Anyone who has experienced a hurricane, tornado, a flood, earthquake, or any other natural disaster, knows how puny the elements on rampage can make a person feel.

Although never winning total victory, the men and women of the Air Weather Service have helped save lives in peace and war, and can hope for a day when world weather can be accurately predicted, measured, and even perhaps one day controlled.

Efforts toward limiting disasters and saving lives, however, are not restricted to the Air Weather Service of MAC. There is another unit of MAC equally as dedicated to saving lives, the group of people we will meet in the next chapter—the men and women of aeromedical airlift.

CHAPTER 5

Flying Ambulances

The crews of the Military Airlift Command's "flying ambulances" have a motto: "Patients are not cargo. Patients are not passengers. Patients are patients."

What the aeromedical airlift crews have to cope with is not precisely normal ambulance or hospital procedure. During emergency or wartime conditions, the medical technicians are trained to load one litter every minute aboard a C-141 Starlifter. And the "room" that each patient is assigned does not depend on income or rank. The flight nurse sets up a "seating" or "litter" chart, according to the tag on the patient indicating the nature of the wound or illness. The tag includes special attention needed and the hospital to which the patient is to be sent.

Emergency word abbreviations are used. MFW means multiple fracture wounds; GSW, gunshot wounds; and, in the chart, added chair space must be alloted to those with casts. Patients with mental illness may be seated between two ambulatory patients, if possible, so that care can be provided as quickly as needed.

The aeromedical aircraft do not routinely carry physicians as crewmembers. Still, ambulance or hospital routine is followed in the giving of needed injections or soothing words. Also, the flying ambulance crews, with their stacked litters, have to be alert for patient problems in flight. These are not just the problems of normal hospital care. They do have hypodermic syringes, bandages, medicines, catheters, pumps—but flying has its own hazards for the patients.

Altitude and turbulence can be problems—although the pilots will go far out of their line of flight to avoid turbulence, which can cause life-endangering air sickness to some patients and even jolt out essential needles and tubes. Also, cabin pressures are not always the same as the ground pressure at takeoff points. Pressure above certain altitudes can cause problems for patients who are having difficulty breathing, even on the ground. And field hospital stitched wounds may hemorrhage if exposed too soon to the low pressures of high-level flight.

Also, the crews are especially trained against sudden decompression in the airplane. Taking masks and carrying oxygen bottles, nurses and medical technicians have to move fast to make sure all the patients have their own individual oxygen supply.

Pilots must be prepared with extra fuel supply to avoid what weather hazards they can, and they receive the latest weather information to find the smoothest flying. They have to make choices, which first must be cleared with air traffic controllers: should they try to climb above turbulence, and how far above can they go and yet avoid endangering a patient?

World War II was the first war in which mass evacuation of battle wounded by air was attempted. Air Transport Command Douglas C-47 ambulance planes were used.

And they have another problem always with them. It is called "block time." Essentially, that means arriving at their destination on time to the minute. The facilities that serve casualties must know when to be there: support vehicles and personnel, hospital readiness, and always, firefighters. Patients on a flying ambulance are seldom in condition to jump or slide readily from a plane in an emergency.

Transporting wounded by airplane from battle areas to hospitals was first thought of back in 1910. Seven years after the Wright brothers' first flight, two young American Army officers, using their own money, constructed and flew an ambulance airplane. The plane was modified for the pilot to sit beside the patient. The two officers, Capt. George Gosman and Lt. Albert Rhoades, went to Washington to recommend that the airplane ambulance be tried. Their presentation was rejected. One newspaper editorial scoffed, "The hazard of being wounded is sufficient without the additional hazard of transportation by airplane!"

But the idea had already been tried successfully during the Franco-Prussian War of 1870. The French launched 66 balloons to airlift 160 sick and wounded out of Paris when it was under Prussian siege. Later, in 1916, the French used a modified Dorand II airplane, designed by a French medical officer, to fly wounded soldiers from the battlefield in Serbia. Two patients were carried behind the cockpit.

While that design proved the feasibility of aeromedical evacuation, World War I (1914–1918) had millions of casualties and scarcely any airplane ambulances. World War I, however, stimulated the development of aeromedical transport. The year the war ended, a Capt. William Ocker and Maj. Nelson Driver at Gerstner Field, Louisiana, converted a JN-4 Jenny biplane into an ambulance by removing the rear cockpit and putting in a stretcher. The Jenny was successfully used in tending casualties in a pilot-training program. Soon, other air ambulances were commissioned for military airfields.

Three years after World War I, the Surgeon General praised the use of airplanes by the U.S. Army patrolling the Mexican border. Cavalry patients, quite possibly "wounded" by their horses, were airlifted to hospitals.

71

A Curtiss JN-4H Jenny equipped as an air ambulance in 1918.

Many modifications of air ambulances appeared in the interwar years that followed, with the increase of cross-country flying. In 1920 a plane called the Cox-Klemin was built especially for air ambulance work, but the most successful of the planes used in air evacuation were a modified Douglas C-1 and a Ford-Stout C-9. There were those in the Air Corps, though, who still did not believe in the concept of using converted transport planes for air ambulance work.

With the start of World War II, aeromedical evacuation became a war necessity. Too many American military personnel were deployed to locations with limited medical facilities, and the old system of surface evacuation proved impractical.

The first planes used for air evacuation of the wounded were transports: C-46 Commandos, C-47 Gooney Birds, C-54B Skymasters. After unloading their cargo, the planes were converted into primitive flying ambulances with stretchers of wounded in some planes stacked four high for the return trip to the States. None of the planes was large enough to carry more than 36 litters.

Formalized training of nurses and enlisted medical technicians to tend the wounded on these first flying ambulances was not begun until January, 1943, at Bowman Field in Louisville, Kentucky. At first there was opposition to the use of female flight nurses, but it was soon proven that the nurses were the most highly qualified medical personnel available. By the end of the war, almost one and a half million casualties were carried by ambulance airplanes from the war zones to the United States. And the first medical evacuation by a new type of flying ambulance—the helicopter—had begun.

First airplane designed especially for air ambulance evacuation was the Cox-Klemin XA-1.

Evacuation of patients by air during World War II was not always aboard ambulance planes, but by cargo planes like this C-47 on an airstrip in the Philippines in 1945.

In 1948, aeromedical airlift became the responsibility of the Military Air Transport Service. Despite cuts in personnel, under MATS a comprehensive aeromedical evacuation system was developed, providing a military member stationed anywhere in the world the best possible medical care.

Yet in the first months of the Korean War, the Eighth Army did not fully utilize air evacuation of the wounded. In part, this was due to the Eighth Army Surgeon who estimated that nine out of ten casualties would be returned to battle. Nevertheless, H-5 and H-19 helicopters saved many critically wounded in Korea, by transporting them from the frontline aid stations. Many of those airlifted patients would not have survived the ten-to-fourteen-hour surface ambulance ride to field hospitals.

In December, 1950, at the time of the Chinese Communist offensive, C-47s evacuated more than 4,000 sick and wounded soldiers. This was the most successful and largest aeromedical evacuation mission during the Korean War. Casualties were carried island to island from Japan to the United States over a period of several days. By the end of the war, aeromedical evacuations totaled 311,673 patients.

A wounded United Nations soldier being carried aboard a C-54 Skymaster in Korea for evacuation to rear-area hospital. C-54s brought in vital war materials and returned with wounded.

In the post-Korean War years, the Douglas C-118, Lockheed C-121, and the Douglas C-124 replaced the C-54 aeromedical fleet. In 1954, the C-131A Samaritan, the first fully pressurized twin-engine transport, could accommodate 40 ambulatory or 27 stretcher patients. The air-conditioned plane could carry almost any type of special medical equipment and was used mainly between hospitals within the United States.

When MATS became the Military Airlift Command in 1966, its aeromedical evacuation mission was firmly established. However, in 1964, at the time of the Gulf of Tonkin incident in Vietnam, the aeromedical airlift system was still at a peacetime level and medical personnel in the Pacific numbered only 60. Medical personnel were hastily brought in from other air commands as well as from the Air Force Reserve and Air National Guard. Six Pacific operating locations were set up and by the end of June, 1968, the number of medical personnel had risen to 1,142.

The C-118 and the Lockheed C-130 Hercules (70–74 litter capacity) moved patients from Vietnam to Okinawa, Japan, and the Philippines. After their conditions stabilized, patients either returned to duty or began the second stage—evacuation to the United States on MAC Boeing C-135 jet Stratolifters. The Army's Bell UH-1 Huey and the Air Rescue's Kaman H-43 Huskies provided helicopter assistance.

In 1965, the C-141 Starlifter revolutionized aeromedical evacuation with its long-range, high-speed jets and larger cabin capacity. The aircraft could carry up to 80 litter patients or 154 troops, in addition to two nurses and three medical technicians. The Starlifters carried cargo in, and then were transformed from freight ships into flying ambulances in as little as forty-five minutes, if necessary.

76

*Americans wounded in Vietnam aboard a C-141 Starlifter en route to the
United States in 1966.*

The C-141 could move patients from Vietnam to the United
States in 17 hours, from Japan to a hospital at Travis AFB, California, in 9 hours. It was possible that a soldier wounded in Vietnam could find himself in a hospital in the States the very same

day. With the new Starlifter flying ambulance in their inventory, air evac crews flew 86,000 battle casualties from Vietnam to the United States from 1966–1971.

In 1969, an ambulance airplane, designed exclusively for USAF aeromedical evacuation missions by McDonnell-Douglas, came into MAC use: the C-9A Nightingale.

For the first time, a flying ambulance provided such specialized services as built-in ramps and stairways, an isolated care area resembling a hospital intensive care unit, electrical outlets permitting the use of cardiac monitors, defibrillator, respirators, and infusion pumps, among other special features. The Nightingales carry medical equipment for almost any medical emergency.

The twin-jet C-9A Nightingale has a range of 2,500 miles and can carry 40 litters or ambulatory patients or combinations of both.

With the C-9A and C-141 working together, a patient can be transported from anywhere in the world to a hospital anywhere in the United States in less than thirty-six hours.

In 1975, the worldwide aeromedical evacuation system was consolidated under MAC, and the 375th Aeromedical Airlift Wing, with its headquarters at Scott AFB, was charged with the management of the entire aeromedical evacuation system. The 375th moves not only military personnel, but their families and other Department of Defense patients to the most medically appropriate hospital. Support is provided by nine Air National Guard and twenty-one Air Force Reserve units.

Located at Scott AFB is the Armed Services Medical Regulating Office (ASMRO). Hospitals, worldwide, call this office to report Department of Defense patients needing aeromedical airlift. At ASMRO, duty controllers (regulators) match patient movement requirements with the capabilities of individual hospitals. Once a suitable hospital has been selected, the information, along with necessary medical information about the patient, is passed on to the Patient Airlift Center (PAC), also located at Scott AFB.

PAC is responsible for managing all aeromedical airlift missions in the continental United States, including Alaska, as well as the Caribbean area and Bermuda. Together with the air evacuation squadrons at Rhein-Main Air Base, Germany, and Clark Air Base, Philippines, the PAC provides 24-hour, seven-day-a-week aeromedical airlift support around the entire world.

The Center at Scott AFB classifies the patients as "urgent" if they must be moved immediately, as "priority" if they must be picked up within 72 hours. Then the Center sets up flight itineraries and relays these requirements to airlift squadrons or operations centers.

Each C-9 Nightingale has a two-man flight crew and a flight mechanic; a third of all the flight crews are Air Force reservists. The flight crews are called the "Frontenders." Before the flying even begins, the required medicines and medical records for each patient must be received and checked by the medical crew aboard the plane. The medical crews are called the "Backenders": normally two flight nurses and three medical technicians.

The patients aboard are often a little surprised at the number of airfields they visit en route to their destination. Patients flying from coast to coast may stay overnight at one of six motel-like "Remain Overnight" medical facilities throughout the United States. Sometimes it may take several days for a "routine" patient to reach his hospital because flight routes must be flexible.

A sudden emergency can cause a rapid change in a routine flight plan. An aeromedical evacuation flight to its home base in Germany was diverted to Beirut, Lebanon. Upon arrival at Beirut, the medical crew was inundated with casualties from the bombing of the U.S. Marine Corps Headquarters in that city. Meanwhile, halfway around the world, a Nightingale en route to Altus AFB, Oklahoma, was diverted to Kentucky to pick up a severely burned young girl. Onboard support systems kept the girl alive during the two-hour flight to San Antonio.

Whether it's a "routine" or an "urgent" flight, the highly qualified crew aboard the Nightingale offers the patients reassurance and care. Patients have always flown in safety. The Aeromedical Airlift Wing has accumulated 432,000 accident-free flying hours with its C-9 fleet. That's equal to flying one airplane 24 hours every day for 50 years without an accident.

Of course, the peacetime flights differ from those of wartime. In peacetime, an estimated 80 percent of patients are ambulatory, not litter patients; in wartime an estimated two-thirds would require fast litter transportation. The peacetime efforts, however, apart from satisfying daily needs, provide added training for personnel, in case of emergencies or conflicts.

The crews of the 375th Aeromedical Airlift Wing fly six to eight missions, move 300–400 patients, land at approximately 60 locations, and service 70–80 medical facilities daily. Because of their continuous training, the crews of the flying ambulances are equally effective in rapidly transporting the injured or ill, including civilians, during disasters or emergency situations.

Following the outbreak of hostilities or a natural disaster requiring urgent aeromedical evacuation, the Tactical Aeromedical Evacuation Subsystem provides a first response capability. The resultant peacetime aeromedical missions of the 375th AAW make an impressive list. The missions have ranged from flying former American prisoners home from Vietnam to airlifting astronauts of the 1974 Skylab 4 crew; from flying American hostages home from Iran after 444 days in captivity to evacuating survivors of commercial aircraft disasters in the Canary Islands and Spain; from flying out survivors of the Jonestown, Guyana, tragedy to evacuating hospital patients in the path of a hurricane.

In 1984 alone, Military Airlift Command aircrews, nurses, and medical technicians provided aeromedical evacuation for 80,217 patients, including military, civilian dependents, retired military personnel, civilians and foreign nationals. The patients were moved on a total of 4,450 missions by C-9As and C-141s. The C-141s were recently remodeled to have a capacity of 100 litter or 210 ambulatory patients.

Statistics can't show the truly human effect of the improved medical care provided by the "flying ambulances," but statistics can give a graphic human picture. Thus, in World War II, four servicemen died out of 100 wounded. In the Korean conflict, the statistics became two fatalities. In the Vietnam fighting, the number dropped to one death out of 100 wounded. Improved techniques and blood transfusions helped, but the people of the MAC worldwide aeromedical airlift system can be proud of the role they play in saving countless number of lives, past, present, and in the future.

CHAPTER 6

Air Force One

"Air Force One" is the radio call sign for an airplane flying the President of the United States. The call is used for air traffic control identification as well as for all communications to and from the President's airplane. While the President has a special airplane for his primary use, that aircraft does not take on the name of Air Force One until the President of the United States is on board. It is his "oval office in the sky." If, instead, the Vice President is the passenger, the radio call sign is "Air Force Two."

Transporting the President of the United States is the responsibility of the 89th Military Airlift Wing of MAC, headquartered at Andrews Air Force Base, Maryland. The Office of the Vice Chief of Staff, United States Air Force, does the scheduling of the various Special Air Missions.

The 89th, though, is not limited to flying the President of the

United States. Their missions also include flying the Vice President, cabinet members, and Congressional delegations as well as dignitaries, kings, queens, and prime ministers of other countries. The 89th Wing flies these top government officials on a worldwide, 24-hour commitment.

Theodore Roosevelt was the first American president to fly in an airplane. Roosevelt had left the presidential office when he took his first flight in 1910 at what was then Kinloch Field, St. Louis, Missouri. The plane in which he flew was a Wright Type B pusher, a primitive wood-and-canvas craft held together by crisscrossing wires, but Roosevelt was typically enthusiastic, calling flying "practical and safe."

For a number of years, American presidents were discouraged from traveling out of the country, and when they did it was by ship. It was not until World War II that President Franklin Roosevelt became the first president to travel by air while in office. In 1943 he flew to Casablanca to confer with British Prime Minister Winston Churchill. Since there was no presidential plane, he was loaned a Navy Boeing 314 Flying Boat, called a Dixie Clipper, for overwater flight, and an Army Air Corps C-54 for overland flight, to make the trip. With change of aircraft, three stopovers were required en route, and the entire trip took 90 hours in the air.

Franklin Roosevelt took only three air trips while he was president, but in 1944 a Douglas C-54 Skymaster, nicknamed the "Sacred Cow," was set aside specifically for presidential service.

In 1947, President Harry Truman, who loved flying, took his air journeys in a Douglas DC-6. Because "Sacred Cow" did not seem a very dignified name for a presidential plane, the DC-6 was painted like an eagle and designated *The Independence*. Along with many technical improvements, including a cruising speed of

The "Sacred Cow," the first aircraft specifically built for presidential use. A DC-4 Skymaster, the plane was fitted with an electrically operated elevator to lift Franklin D. Roosevelt's wheelchair.

315 miles an hour, four 2,100-horsepower engines, automatic pressurization, and air conditioning, the President for the first time could keep in touch with Washington with a radio teletype transmission system. For the first time, also, the interior of *The Independence* was arranged so that the President could work on board, comfortably. He could even, as President Truman once mischievously did, have the pilot buzz the White House.

President Dwight Eisenhower, who was the first president to have been a pilot in his own right, also liked flying. He had two

The Independence, *Harry Truman's aircraft, was painted to resemble a giant American eagle. The three windows to the rear mark the President's suite.*

new modernized aircraft: a Lockheed Constellation and Lockheed Super-Constellation, designated *Columbine II* and *Columbine III.* (*Columbine I* was the Constellation assigned to President Eisenhower when he served as Commander in Chief of Allied Forces during World War II.) The Constellations were named after the blue columbine, the official flower of Mamie Eisenhower's home state, Colorado. A confusion of radio frequencies on *Columbine II* brought about a different control system and adoption of the radio call sign "Air Force One."

In 1959, a jet transport Boeing 707 became the presidential aircraft. The Boeing 707 offered more speed, and the President of the United States in his propeller-driven Constellation would no longer be outshone by Russian dignitaries flying in their new turbojet aircraft.

The most history-making presidential plane was Special Air Missions (SAM) Aircraft 26000, a new Boeing 707, which joined the presidential fleet of President John F. Kennedy in 1962. Luxuriously appointed inside, the words UNITED STATES OF AMERICA were painted on both sides of the fuselage and the American flag was painted on the tail. In May, 1963, on a trip to Moscow with a United States delegation, Air Force One, which had become the official designation for the presidential plane, set 14 speed records, including flying from Washington to Moscow in a record-breaking 8 hours, 38 minutes, 42 seconds.

But this same SAM 26000 made another kind of history as Air Force One when it flew President Kennedy to Dallas, Texas, on November 22, 1963, and flew his body back to Washington, D.C., following his assassination.

Before the trip back, on board this same aircraft at Love Field, Dallas, Lyndon B. Johnson was sworn into office as the 36th president of the United States. He was the first president to be sworn into office aboard an airplane. President Johnson, who it was said ran the SAM 26000 "like a range boss on a cattle drive," increased the number of seats aboard the plane to allow for more passenger space. He also had the seats changed so that passengers flew backwards, facing him, and his special chair and kidney-shaped conference table. His last trip aboard Air Force One was after his state funeral in Washington, D.C., on January 24, 1973, returning his body to Texas.

A remodeled Aircraft 26000 was later used extensively by President Richard M. Nixon during the first four years of his administration. His trips included an around-the-world flight in 1969, a trip

The most famous photograph ever taken aboard a presidential airplane was the swearing-in of Lyndon B. Johnson at Dallas' Love Field on November 22, 1963. The President is flanked by Lady Bird Johnson and just-widowed Jacqueline Kennedy.

to the People's Republic of China in February, 1972, and to Russia in May of that same year. Modern presidents, unlike earlier ones, are a much-traveled group, but President Nixon visited more countries and traveled more miles outside the United States than any other president.

Even after Aircraft 26000 was relegated to being only an "alternate" aircraft for presidential use, the British Queen Elizabeth flew on it through her travels in the western United States in March, 1983.

The replacement Air Force One, Aircraft 27000, became the presidential plane in 1972. It has been used by Presidents Richard Nixon, Gerald Ford, Jimmy Carter, and Ronald Reagan. It is a specially configured Boeing 707-353B, with the Air Force designation C-137C. This new Air Force One was specifically equipped to meet the president's needs, including secure global communications. The passenger cabins are partitioned into several sections with separate presidential quarters and staff/office compartments. There is also limited seating for other passengers, including news media representatives.

Air Force One, Number 27000, over Mount Vernon, home of the first president. The Boeing 707 is specially outfitted to meet the president's needs.

In addition to the two Boeing 707s, the 89th Military Airlift Wing of MAC has in its inventory five C-137s, six C-135s, four C-140 Jetstars, three C-20A Gulfstream IIIs, three C-9s, two C-12s, one C-6, and a helicopter squadron of UH-1Ns and CH-3Es. The aircraft are used by the 89th's six squadrons and one operational detachment for their other diverse duties, ranging from providing C-12 flying training for U.S. Embassy attache pilots to using their helicopters for high-priority airlift and, on emergency, medical evacuation missions in the Washington area.

President Ronald Reagan continues to use as his presidential plane Aircraft 27000, and calls it his "private turf," but the next Air Force One is already in the planning stage and will undoubtedly be a larger and wider aircraft. Although the two special presidential planes—Aircraft 27000 and backup Aircraft 26000—are probably the most pampered, looked-after aircraft in the world, one is twenty years old and one is ten. The future Air Force One will have more space for improved communications, staff, conference and medical facilities.

No matter what type of plane the President flies, one thing will not change. The ever-present attache case, which contains the codes for ordering a nuclear strike, will be on board. It accompanies the President everywhere, carried by a special military aide, and it will be on board as soon as the President boards.

The mission of flying dignitaries aboard the diversified aircraft of the presidential fleet requires a very special organization. Not only must the very important passengers be carried safely to their destinations, but rigid time schedules must be maintained so that the passengers will arrive punctually at their various destinations. Often the time schedules for the VIP flights are arranged a month in advance with little knowledge of the weather prospects. Never-

President Ronald Reagan and wife, Nancy, in front of Air Force One on a visit to Pacific Missile Test Center, Point Mugu, California, in 1981.

theless, on one mission taking a high-level government official to more than thirty en route stops around the world, the presidential aircraft was off schedule less than a minute!

Safety and security precautions for the President and Air Force One also require special attention. The crew of Air Force One test-fly and examine the plane constantly to make sure everything is functioning properly. Secret Service men and women and Air Police Security police guard Air Force One on the ground.

While in the air, the flight of the presidential plane is monitored electronically by the National Security Agency as well as the

National Military Command Center. A special weather team provides the latest weather information anywhere in the world to the pilot of the aircraft. Planes of MAC's Air Rescue Service accompany Air Force One on overseas flights, to assist in case the presidential plane should have to ditch over the ocean. If necessary, Air Force One can immediately order a fighter plane escort.

The crews who fly the presidential planes are handpicked and must meet rigid requirements. Pilots must have at least 3,000 hours flying time, 200 hours of which must be as an instructor pilot. Navigators must possess a minimum of 2,000 hours flying time over worldwide routes. Flight engineers and radio operators have an average of 14 years experience.

Air passenger specialists who act as flight stewards aboard the plane must be experts in emergency procedures, and consummate diplomats in handling their important passengers, as well as skilled chefs. Maintenance airmen must meet both the tough experience requirements of the Air Force and the Federal Aviation Agency.

The highly qualified personnel who maintain and fly the aircraft of the 89th Military Airlift Wing have been awarded the Air Force Outstanding Unit Award seven times. They fly with the motto "Pursuit of Perfection," aware that their job is to carry some of the most important people in the world—safely. That they have carried out that job is clearly on record. Their more than 700,000 flying hours have been accident-free.

CHAPTER 7

Air Commandos

It was night when a MAC MC-130 with a planeload of Army Rangers reached Grenada, the first aircraft to arrive over Point Salines airfield. As the Rangers parachuted into the darkness, the MC-130 was suddenly spotlighted by a searchlight from the airfield and the plane came under intense antiaircraft fire.

Flying at 500 feet and only 120 knots, the unarmed plane—and those following it—was dangerously vulnerable to the enemy fire. But help was on the way. An AC-130 Spectre Gunship from MAC's 1st Special Operations wing, circling at a high altitude above the field, made a pass over the enemy antiaircraft guns on the field, effectively silencing them.

The success of the Special Operations Forces in working smoothly with the airlift forces during the Grenada operation is particularly noteworthy because the Special Operations Forces

(SOF) had been consolidated with the Military Airlift Command only eight months before.

Although MAC's rescue and recovery service has similar missions, training, and equipment as the Special Operations Forces, SOF had been a separate organization of the Air Force until it became a part of MAC in March, 1983. A new numbered Air Force, the 23rd—called the combat arm of MAC—was set up at that time. The 23rd includes, among other worldwide missions, combat rescue and recovery, aeromedical evacuation, weather reconnaissance and aerial sampling, security support for intercontinental ballistic missiles—and special operations.

Special Operations Forces, also known as Air Commandos, specialize in "unconventional warfare, collective security, counterterrorist operations, psychological operations, and civil affairs measures."

To explain the Air Commandos—their aircraft, actions, and purpose—we must first go back to their beginning in World War II in the China-Burma-India theater of operations. Survival behind enemy lines there for British troops under the command of General Orde Wingate, who made daring, long-range raids into Japanese-held Burma, would have been even more difficult without the support of the American 1st Air Commando Group.

In one of the most unusual operations of that war, the Air Commandos, using C-47s towing gliders, transported Wingate's raiders and their equipment over a 7,000-foot mountain range, releasing their burdens 100 miles behind Japanese lines. Then they set up a remarkable air shuttle operation, airlifting in supplies for the British troops to rough airstrips hacked out of the jungle, and carried out Wingate's casualties.

Disbanded after the war, the Air Commandos and the idea of a

Some of the most difficult cargo to be infiltrated behind enemy lines were pack mules airlifted to Wingate's Raiders in Burma by Air Commandos during World War II.

Special Operations Force lay dormant until President John F. Kennedy called on the military to train its forces to fight in a new type of unconventional warfare, "a war by guerrillas, subversives, insurgents, assassins, war by ambush instead of by combat." The Air Force responded by activating at Hurlburt Field, Florida, in 1961, the 1st Air Commando Wing, eventually known as the 1st Special Operations Wing.

The types of airplanes of this Wing, then and now, were as special as those who used them. Thus, during the Vietnam War, the aircraft used by the Air Commando units were often vintage models, propeller aircraft, such as T-23s, C-47s, B-26s, and A-1s, appearing out of place in the jet age. But as one Air Commando commented, "Our planes may be obsolete and unsophisticated but they can do our kind of job!"

Their job? Picture at midnight several shadowy figures waiting, hidden near the tree line of a jungle in North Vietnam. The powerful roar of turboprop engines suddenly shakes the pitch-black night. An almost invisible MC-130E "Blackbird" crests a nearby ridge, ejects a package, makes a descending turn, then disappears again into the darkness. The bundle of ammunition and supplies

Special Operations Forces in Vietnam often used conventional transport planes in unconventional ways. The modified C-47, armed with guns at cargo doors and windows, was nicknamed "Puff the Magic Dragon."

is retrieved quickly from the drop zone by the dark-clothed Special Forces team. They then bury the parachute and container and disappear quickly and silently back into the jungle.

Such a night action, supplying Special Operations teams behind enemy lines, was a standard USAF Special Operations mission during the Vietnam War, involving "precise timing and coordination deep behind enemy lines, working under the cover of darkness." But the Air Force Special Operations job ranged also from dropping psychological warfare leaflets to setting up and supplying isolated Special Forces camps and sending Long-Range Reconnaissance patrols into enemy territory.

To defend the Special Forces camps from nighttime enemy attacks, the Special Air Warfare/Air Commando forces successfully developed the workhorse C-47 transport into a flareship, so that it could operate at night. Nicknamed "Puff the Magic Dragon," the aircraft was also armed with 30-caliber machine guns and 7.62-mm miniguns jutting from the passenger windows and cargo doors. The guns could be both sighted and fired by the pilot while the aircraft was in a left bank.

Another aircraft modified during the Vietnam War for Special Operations activities was the Hercules C-130, which became the AC-130 Spectre Gunship and in another modification, the MC-130E Combat Talon, known as "Blackbird."

The armed Spectre Gunship was equipped with sophisticated sensors with which to locate targets. An onboard computer automatically figured gun alignment and aimed the guns with deadly accuracy. The Combat Talon was designed for long-range penetration of hostile areas so that it could bring in, resupply, and remove Special Operations teams working behind enemy lines. Special terrain-following radar allowed the Combat Talon to fly

The Combat Talon performs long-range, night, adverse-weather, low-level infiltration, resupply, and other missions.

over mountainous terrain at very low altitudes, and the plane's inertial navigation system and mapping radar made possible accurate airdrops on unmarked drop zones, day or night.

An interesting modification of the Combat Talon was the addition of the Fulton (Skyhook) Recovery System to rescue downed crewmembers. The aircraft was outfitted with a yoke on the nose resembling the feelers of a large beetle. The person on the ground was dropped a kit containing a harness suit, a lift line, a balloon, and two containers of helium. The individual dons the suit, makes sure the lift line is attached to the suit and balloon, then inflates the balloon with the helium and lets it fly. The aircraft was maneuvered to intercept the lift line with the yoke, attaching the line to the aircraft by a special lock. The crewmember was then pulled smoothly into the air and winched on board the Combat Talon.

Armed helicopters such as the HH-53H Pave Low and the UH-IN Twin Huey were also flown by Special Operations personnel. Equipped with specialized electronic equipment, such as terrain following/avoidance radar, the helicopters could provide close-in fire support as well as removing Special Operations teams from behind enemy lines in all kinds of weather.

Even after the end of the Vietnam War, the need for a Special Operations capability remained. Suddenly erupting international

Two men being winched into open cargo door of aircraft equipped with Fulton Aerial Recovery System.

turmoil and terrorist threats that reached around the world required a small, flexible, well-trained force able to protect American interests and citizens abroad.

Many of the Special Operations aircraft used in Vietnam and in Grenada are still in use today. However, both the MC-130 Combat Talon and the AC-130 Spectre Gunship, as well as the combat rescue and Special Operations helicopters, are being augmented and updated with improved navigation equipment and special avionics capabilities. An experimental aircraft, the JVX, a fixed-wing turboprop aircraft with vertical lift (helicopter) capability, is on the drawing board.

Night-vision goggles which can amplify objects 40,000 times their normal brightness at night, producing an eerie black-and-green image, used by combat rescue teams, are now being used by Special Operations as well.

And what of the Special Operations men and women? Well, like their aircraft, they, too, are a special breed. Some—by no means all—are trained at the USAF Special Operations School at Hurlburt Field, Florida, where the curriculum covers a diverse range of subjects from unconventional warfare, psychological operations, insurgent operations and international terrorism to foreign internal defense.

Many are taught skills not found in a classroom. A Special Operations Combat Control Team, which does everything from establishing drop zones behind enemy lines to guiding aircraft to targets in remote areas, is trained in parachute jumping, radio main-

Combat Control Team from 1st Special Operations Wing rappels from a hovering Army UH-1H Iroquois helicopter during training.

100

tenance, scuba diving, forward air control operations, and much more.

The members of the MAC Special Operations Forces are prepared for combat in the field, but know that their most critical role is supporting Army, Navy, and allied special operations teams operating behind enemy lines.

First and foremost, the MAC Special Operations Forces are airlifters who "anytime, any place, get them there, resupply them, and get them out!"

CHAPTER 8

Airdrop or Airland– Anything, Anywhere

A sandy field is covered with airdropped armored reconnaissance vehicles, jeeps, and artillery. The scattered pieces of equipment are still palletized and rigged for airdrop. Soon C-130 transports appear overhead, approaching in a long staggered column. Over the drop zone the planes release their cargo: paratroopers of the 82nd Airborne Division.

Canopies inflate and quickly descend. Once on the ground and out of their chutes, the troopers, wearing camouflage combat fatigues and carrying field packs and automatic rifles, rush to the field. Within minutes, the paratroopers have the heavy equipment derigged. Engines roar to life, howitzers are in position, and the jeeps move toward the enemy objective.

The above joint military exercise took place at Normandy Drop Zone at Fort Bragg, North Carolina, a scenario that involved U.S.

Army paratroopers but which could not have taken place without the close support of the Military Airlift Command.

Training jump school is at Fort Benning, Georgia, but all air-drop wings take part in parachute troop training. The training requires five training jumps from C-130 and C-141 transports, including a night jump and at least one in full combat gear. Thus, at Pope AFB, almost in the middle of the Army's Fort Bragg, the 317th Tactical Airlift Wing of MAC is responsible for training drops from C-130s of 130,000–140,000 troops a year.

Even after a paratrooper receives his wings, the intense training continues, making sure that the 82nd, the spearhead of a Rapid Deployment Force, is ready to be airlifted to fight anywhere in

Troops wait to board a C-141B Starlifter at Pope AFB at Fort Bragg. Helping in the training of paratroopers is part of MAC's airdrop mission.

the world. One of the 82nd's three battalions is always on standby alert, able to deploy within twenty-four hours' notice. Within the battalion, one company is on two-hour alert, ready to move out immediately.

Airborne troops can carry only a limited amount of supplies and ammunition with them. Once on the ground, they must be resupplied, usually again by airlift. If the transport planes can't airland—that is, off-load their cargoes on the ground—because of weather, enemy action, or lack of suitable airstrips, then the supplies must be airdropped. Airdropping of supplies, not only to airborne troops but to all combat troops, can present enormous problems when the equipment that must be dropped ranges from Sheridan tanks to fragile communications equipment.

First, aircraft with cavernous cargo spaces must be found, large enough to carry anything from a howitzer to a helicopter. Second, with these heavy payloads, the aircraft must still be able to take off and have sufficient range and airspeed to reach its destination. Third, if the cargo can't be off-loaded on the ground, then a way must be found to safely deliver the equipment by airdrop, as precisely as possible within a designated drop zone.

During World War II, the Air Transport Command, predecessor of MAC, used some of the largest transports for military cargo: C-46s, C-47s, and C-54s. The C-47, which has probably hauled more freight than any other aircraft in history, could carry only 6,000 pounds on long hops. The C-54 Skymaster, the largest mass-produced four-engine cargo transport of the war, carried a top payload of 14,000 pounds of freight.

Early in the war, the delivery system of the cargo on these planes which couldn't be airlanded was simple. The cargo was rigged

A jeep being loaded aboard a C-46A transport plane. The C-46 was the largest twin-engine cargo transport of World War II, and was also used in Korea and Vietnam.

with parachutes which were then airdropped as the plane made low passes at the ground. Accuracy was a problem. Thousands of pounds of precious supplies were lost when they missed a tiny clearing and were swallowed up by the jungle. In addition, it took pass after pass to complete an airdrop, the aircraft flying at stalling speed only 150 feet above hostile ground, all the while exposing the flight crews to enemy ground fire.

Every cargo plane had its unsung hero, the "kicker." His job

106

was to drag heavy bundles of supplies to the open door of the plane, hang onto anything solid his hands could reach, then kick with all his might, pushing the heavy parcels out to troops eagerly awaiting the supplies on the ground. Sometimes the "kicker" would lose his grip and follow the bundle down, or be wounded by enemy fire as he stood in the open cargo door.

Toward the end of the war, the C-47s were fitted with a conveyor, occupying the left half of the cabin from aft of the pilot's compartment to the cargo door. An electric motor drove the belt, which could support about 4,000 pounds on its 22 inch-wide endless belt. When the parachute-rigged containers reached the launching platform at the cargo door, static lines opened the chutes as a mechanism nudged the packages out the door. With the conveyor, a plane could drop 500 gallons of gasoline in seven seconds and in a space of 300 yards.

The first U.S. military transport was also developed during World War II, the Fairchild C-82 Packet. Previous military transports had been off-the-shelf civilian aircraft, modified for military airlift. The C-82 was replaced by the Fairchild C-119 Flying Boxcar, the biggest and most powerful cargo transport of its time. Wheeled equipment was loaded on the plane via ramps through the hinged rear cargo door. The clamshell cargo doors at the rear would also allow two rows of paratroopers to jump in unison. An electrically operated monorail dropped twenty 500-pound paracans through a hatch in the bottom of the fuselage. The C-119 had a payload of 20,000 pounds.

The Flying Boxcar proved its worth in Korea when the 1st Marine Division was cut off by the Chinese Communists at Chosin reservoir. Supplies dropped from C-119s kept the Marines alive and fighting for ten days. Then, when the Marines broke out of

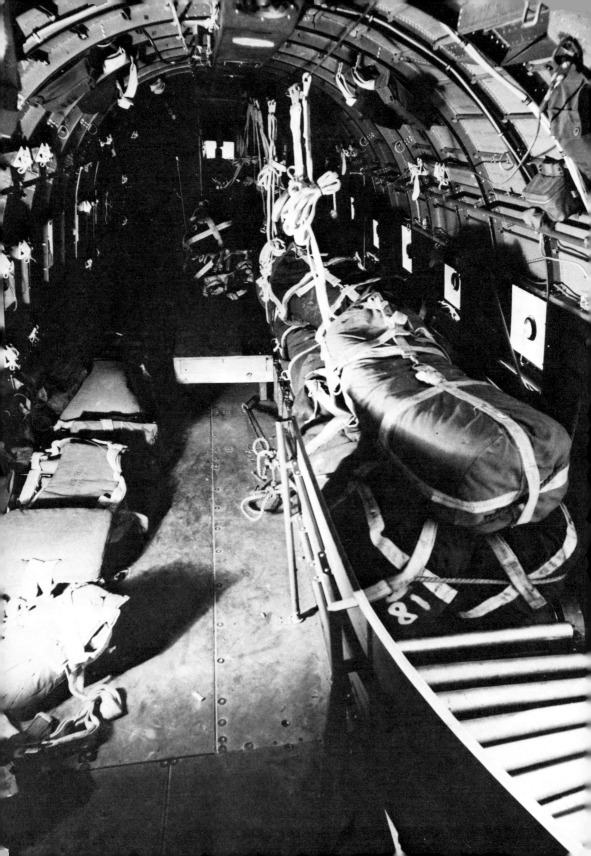

the trap, the parts of a 32-ton bridge were airdropped. The bridge was used to help the Marines cross an impassable gorge to reach the fleet at Hungnam. It was the first time in history that an entire bridge was airdropped.

The Douglas C-124 Globemaster II was also a well-known transport in the Korean conflict. Dubbed "Old Shakey," the C-124 had clamshell loading doors and hydraulic ramps in the nose and an elevator under the aft fuselage. The Globemaster was easily able to handle bulky cargo such as tanks, field guns, bulldozers, and trucks. Its interior could be double-decked to carry 220 fully equipped combat troops; its payload was 74,000 pounds. The C-124's range of 2,300–4,000 miles made it a prime hauler in strategic airlift.

Even during the Korean War, however, the Air Force realized that the piston-driven strategic airlift transports, the C-54 Skymaster and the C-124 Globemaster, lacked the range and cargo capability for supplying a war halfway across the world from the United States. And within the Korean combat theater of operations, the tactical airlift transports, the twin-boom C-119, the hard-working C-47 and C-46, were not large enough to handle the heavy support equipment needed by the frontline forces. Lockheed-Georgia accepted the challenge to develop a new transport plane, and the legendary turboprop C-130 Hercules was born.

Especially developed for takeoffs and landings in as little as 2,000 feet on rough dirt strips, the "Herk" became the prime tactical transport for airdropping or airlanding men and equipment into hostile areas.

An early, factory-style conveyor belt used for airdropping cargo from Air Transport Command transport planes during World War II.

109

C-130D Hercules modified with wheel-ski landing gear for delivering cargo in the Arctic and to units along the Distant Early Warning line.

Up to six pallets of cargo, or 42,000 pounds could be loaded onto the C-130 through a hydraulically operated main loading door and ramp in the rear of the aircraft. The ramp could be lowered for loading and unloading wheeled vehicles. Rollers in the floor of the cargo compartment made the handling of cargo pallets quick and easy. Sixty-four fully equipped paratroops could exit the aircraft through two doors, one on each side of the aircraft behind the landing gear fairings.

With the coming of the jet age, even the legendary "Herk" was not large enough, nor fast enough, to handle airlift requirements in a modern, conventional war. What was needed was a fleet of large jet-propelled transport planes, aircraft capable of picking up an entire Army division with all its equipment and airlifting it overseas in a matter of hours, ready to fight. This new combat cargo plane would have to be easy to load and unload, built to go in low under enemy radar, drop men or supplies and pull up fast and be gone before enemy guns could be brought to bear.

Shortly after the war began in Vietnam, such a plane entered MAC's inventory—the Lockheed C-141 Starlifter. The Starlifter could be used to lift combat forces over long distances, could carry 69,925 pounds or 155 paratroops, and had the capability to airdrop or airland cargo and troops. In the 1980s, the Starlifters were "stretched." A section was inserted in front and behind the wing, lengthening the plane by 23 feet 4 inches, and increasing its cargo volume capability but decreasing its cargo-fuel weight use. At the same time the capability to refuel in flight was added.

In 1969, the C-5 Galaxy was added to MAC's cargo fleet. The C-5 is the world's largest aircraft, almost as long as a football field and as high as a six-story building. The cargo compartment is about the width of an eight-lane bowling alley and a little longer than the distance of the Wright brothers' first flight!

Able to be refueled in the air, the C-5 can carry outsize cargo such as three CH-47 Chinook helicopters, or two 59-ton tanks, or a 74-ton mobile bridge at intercontinental ranges and jet speeds.

Cargo can be loaded and off-loaded simultaneously at the front and rear cargo openings of the C-5. The plane is also capable of "kneeling down" to handle loading directly from truck bed levels. Full-width drive-on ramps at each end are ready for loading double rows of vehicles, pallets, or containers. Troops may be carried on a second story or upper-deck compartment, immediately available to drive vehicles off the plane for airland delivery.

Although MAC continues to use the C-130 Hercules in tactical airlift, the C-141 and C-5 can be teamed to carry fully equipped and combat-ready military units to any point in the world on short-term notice. The aircraft can then provide full field support to maintain the fighting force for unlimited periods.

The C-130, C-141, and C-5 have come a long way from the days when "kickers" were used to airdrop cargo. Today, there are many improved methods of delivering supplies and equipment.

AIRLAND—Preferred delivery by landing aircraft at permanent or temporary airfield and off-loading troops or supplies. Cargo doors are lowered and the pallets of cargo can be off-loaded by conveyors while airplane is taxiing, minimizing ground time in case of enemy action.

FREE FALL—Oldest form of airdrop without a parachute and limited to indestructible items such as bales of hay or wire. Aircraft makes low pass and then a shallow climb while the cargo is rolled out of the aircraft.

CONTAINER DELIVERY SYSTEM—CDS uses the force of gravity to pull bundles of supplies from the aircraft. When the specially designed containers, weighing up to 2,000 pounds, are out of aircraft, parachutes inflate and lower them to the ground. Ammunition and supplies can be delivered accurately within a 100- to 400-meter area. Shock-absorbing materials deliver fragile items safely.

LOW-ALTITUDE PARACHUTE EXTRACTION SYSTEM—LAPES uses parachutes, not to lower the load to the ground but to pull it from the aircraft as it dives to a point five to ten feet above the ground, and levels off. A small drogue parachute deploys up to three larger

This 57-ton marine propulsion-reduction gear for the Navy is shipped aboard a C-5 Galaxy, the only aircraft capable of transporting a load of such weight and size.

A C-130 Hercules airdropping cargo in the low-altitude parachute extraction system (LAPES).

parachutes from rear of the aircraft. These chutes pull out as many as three platforms loaded with up to 37,000 pounds of cargo/ equipment. The whole package slides to a halt on the ground in approximately 100 yards. LAPES is used for delivery of bulldozers, artillery pieces, fuel and water bladders.

HIGH-ALTITUDE AIRDROP RESUPPLY SYSTEM—HAARS is used for cargo drops of 2,000 pounds up to altitudes of 10,000 feet. This also allows delivery of paratroops qualified in high-altitude, low-opening parachute jumping.

The crewman aboard the cargo plane responsible for the safe storage and delivery of cargo is the loadmaster, the modern-day

114

relative of the "kicker." Not only must the loadmaster be able to handle and deliver unusual cargoes, ranging from missiles to once even a whale, but he must also know how to stow—safely—dangerous cargo such as explosives, acids, and gases.

But the MAC unit that most crewmembers rely on is the Air Force Combat Control Team. Composed of jump-qualified specialists, they are among the first to jump on an objective. Once on the ground, they set up vital navigation and communication aids to guide incoming aircraft to the drop zone.

Combat Control Teams were organized in World War II. Originally called Pathfinders, they were deployed before the first wave of airborne troops, and at times had to engage in combat with the enemy before being able to complete their primary mission— providing visual guidance for incoming aircraft.

Today, Combat Control Teams use much more sophisticated radio equipment to communicate with incoming aircraft and the training for combat controllers' jobs goes far beyond learning to establish drop and landing zones and controlling flight operations.

The red berets, as Combat Controllers are called because of their distinctive headgear, are trained to be parachutists, to be proficient in scuba operations, to rappel from helicopters, to be skilled in techniques of survival, radio communications, vehicle maintenance, meteorology, patrol, ambush, and demolition. Most important, they must have the ability to make quick, logical decisions. Many lives, including those of paratroopers, depend on their air traffic decisions.

At one time it was thought that helicopters could take the place of airborne soldiers. However, no one has yet found a substitute for the airborne's capability of projecting a large military force by aircraft over long distances to seize and hold terrain.

The USAF Airlift Center of MAC, at Pope AFB, is constantly developing and testing new and better airlift and airdrop techniques. A new transport aircraft for MAC is also in the planning stage. Even the C-5 is no longer able to airlift the Army's outsize combat equipment developed in the last years. The proposed new C-17 aircraft will perform the full spectrum of airlift missions. It will be capable of carrying all classes of cargo over intercontinental distances directly to the final destination. It will be able to airland or airdrop a payload and perform low-altitude parachute extractions. In addition, the C-17 will have the capability of operating into small airfields previously restricted to C-130s.

From "flying the Hump" in World War II to Grenada, the Military Airlift Command has proven and will continue to prove the capability of airlift to deliver anything, anywhere, anytime!

CHAPTER 9

Behind the Scenes
at MAC

"Keep 'em flying" was a popular slogan for the United States Air Force during World War II. The aircraft of the Military Airlift Command is "kept flying" not just by pilots, copilots, and navigators but by a great many military and civilian men and women working behind the scenes.

Cargo handlers, maintenance specialists, computer experts, researchers, weathermen, aeromedical technicians, and many others are needed to keep things moving. MAC's goal is to deliver anything, anywhere, anytime. It is the largest, busiest, and most far-ranging aerial cargo carrier in the world, and it takes a lot of personnel to do that.

The photographs that follow are only a few glances "behind the scenes" at the wide variety of jobs performed by support services in the Military Airlift Command.

MAC technicians inspecting engine of a C-141.

118

Instrument specialist working on helicopter panel controls.

Specialist processes passengers at the air passenger terminal for MAC flights.

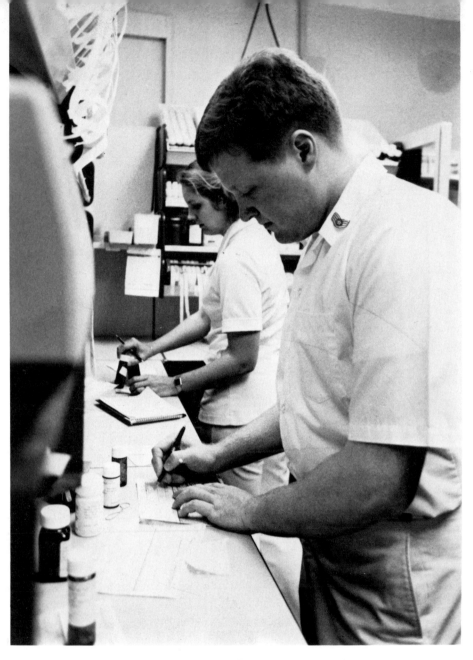

Trained work at a pharmacy is also a job for MAC personnel.

MAC C-130 loadmaster at work.

120

Security policeman training a working military dog.

MAC civil engineers in hard hats inspecting construction.

122

An Aerospace Audiovisual Service motion picture photographer in action during an Air Force exercise. MAC's AAVS provides video, motion picture, and still photographs for all activities of the U.S. Air Force.

124

CHAPTER 10

MAC'S Planes

There have been at least 100 different transport planes developed since the Fokker T-2 Transport, the Army's pioneer monoplane passenger carrier, flew the first nonstop cross-country flight in 1923. The following is a selection of some of the outstanding "workhorse" transports which, through continued use over a period of years, have proven invaluable to strategic and tactical airlift in peace and war.

Almost all have appeared in different models and performed a wide variety of functions. The C-47 has been a transport, a gunship, and an ambulance.

Beautifully restored historic transports may be seen at the National Air and Space Museum, Smithsonian Institution, Washington, D.C., and at the U.S. Air Force Museum, Wright-Patterson Air Force Base, Dayton, Ohio.

C-47 Gooney Bird

DOUGLAS C-47 Perhaps no other transport plane has won the devotion and loyalty of more military pilots than the C-47 Gooney Bird. A low-wing monoplane with a standard tail section and conventional landing gear with retractable front wheels, the C-47 was developed in 1942 from the Douglas DC-3 and has flown in every war from World War II to Vietnam.

SPECIFICATIONS (C-47A)

CONTRACTOR: Douglas Aircraft Company
POWER PLANT: Two Pratt & Whitney R-1830-92 14-cylinder radial-piston engines, 1,200 hp each
DIMENSIONS: Wingspan, 95 ft; Length, 64 ft, 5 in
NORMAL MAX TAKEOFF WEIGHT: 26,000 lb
MAX LEVEL SPEED AT 8,500 FT: 229 mph
NORMAL RANGE: 1,500 miles
LOAD: 28 troops or 18 stretchers and attendants, or 6,000 lb
CREW: Three

C-46 Commando

CURTISS C-46   The C-46 Commando also has the distinction of having fought in three wars. A low-wing monoplane with retractable landing gear and a standard tail section, the C-46 was first delivered in 1942. The largest twin-engine cargo transport of World War II, the C-46, like the C-47, was produced with many variations and has been called the most modified aircraft of World War II.

SPECIFICATIONS (C-46A)

CONTRACTOR: Curtiss-Wright Corporation
POWER PLANT: Two Pratt & Whitney R-2800-51 18-cylinder radial-piston engines, 2,000 hp each
DIMENSIONS: Wingspan, 108 ft, 1 in; Length, 76 ft, 4 in
NORMAL MAX TAKEOFF WEIGHT: 45,000 lb
MAX LEVEL SPEED AT 15,000 FT: 269 mph
NORMAL RANGE: 1,200 miles
LOAD: 50 fully equipped troops, or up to 33 stretchers and four attendants, or 16,000 lb cargo
CREW: Three to four

C-54 Skymaster

DOUGLAS C-54   The Skymaster was the first four-engine cargo transport in military airlift history. Developed from the DC-4 civilian airliner in 1942, the C-54 was a low-wing monoplane with retractable tricycle landing gear. The most-used aircraft during the Berlin airlift, the C-54 also has the dubious distinction of being the first plane destroyed in the Korean War. The C-54C was the first plane to be specially modified as a presidential aircraft.

SPECIFICATIONS (C-54A)

CONTRACTOR: Douglas Aircraft Company
POWER PLANT: Four Pratt & Whitney R-2000-7 Twin Wasp 14-cylinder radial air-cooled engines, 1,290 hp each
DIMENSIONS: Wingspan, 117 ft, 6 in; Length, 93 ft, 11 in
MAX TAKEOFF WEIGHT: 62,600 lb
MAX SPEED: 265 mph
RANGE: 3,900 miles
LOAD: 50 equipped troops or freight load of 14,000 lb
CREW: Six

128

C-118 Liftmaster

DOUGLAS C-118   The Liftmaster, derived from the Douglas DC-6, was a low-wing monoplane with retractable tricycle landing gear, and began to arrive at Air Transport Command bases in late 1946. A C-118 was used as a presidential transport by President Truman and C-118s were used for aeromedical evacuation during the Vietnam War.

SPECIFICATIONS (C-118)

CONTRACTOR: Douglas Aircraft Company

POWER PLANT: Four Pratt & Whitney R-2800-52W 18-cylinder radial-piston engines, 2,500 hp each

DIMENSIONS: Wingspan, 117 ft, 6 in; Length, 105 ft, 7 in

MAX TAKEOFF WEIGHT: 107,000 lb

MAX SPEED AT 18,000 FT: 360 mph

NORMAL RANGE: 3,350 miles

LOAD: 76 troops, or 60 stretchers and six attendants, or 27,000 lb

CREW: Five

129

C-119 *Flying Boxcar*

FAIRCHILD C-119 In 1949 the Flying Boxcar, developed from the C-82 Packet, was the largest and most-powerful cargo transport of its time. A twin-boom, high-wing monoplane, its twin tailfins were connected by a long tail plane. Large clamshell rear cargo doors were embedded with a standard door, allowing two rows of paratroops to jump in unison. The C-119 served in Korea and Vietnam. A modified version, the AC-119G gunship, with the code name "Spooky," had four 7.62-mm mini-guns in the fuselage.

SPECIFICATIONS (C-119K)

CONTRACTOR: Fairchild Aircraft Corporation

POWER PLANT: Two Wright R-3350-999-TC 18EA2 piston engines, 3,700 hp each, and two General Electric J85-GE-17 auxiliary turbojet engines, 2,850 hp each

DIMENSIONS: Wingspan, 109 ft, 3 in; Length, 86 ft, 6 in

MAX TAKEOFF WEIGHT: 77,000 lb

MAX LEVEL SPEED AT 10,000 FT: 243 mph

RANGE WITH MAX PAYLOAD: 990 miles, 20,000 lb

LOAD: 62 fully equipped troops or 40 equipped paratroops

CREW: Four

130

C-124 Globemaster II

DOUGLAS C-124 The Globemaster II was developed from the earlier Douglas C-74 and delivered to Air Force bases in 1950. The aircraft featured clamshell loading doors and hydraulic ramps in the nose and an elevator under the aft fuselage. The Globemaster was able to handle bulky cargo such as tanks and bulldozers, as well as serving as an airborne ambulance. The C-124C served in Korea and Vietnam.

SPECIFICATIONS (C-124C)

CONTRACTOR: Douglas Aircraft Company

POWER PLANT: Four Pratt & Whitney R-4360-63A 28-cylinder radial-piston engines, 3,800 hp each

DIMENSIONS: Wingspan, 174 ft, 2 in; Length, 130 ft, 5 in

MAX TAKEOFF WEIGHT: 194,500 lb

MAX LEVEL SPEED AT 20,000 FT: 304 mph

RANGE WITH 26,375-LB PAYLOAD: 4,030 miles

LOAD: 200 troops or 127 stretchers, 52 ambulatories and 15 attendants, or up to 68,500 lb cargo

CREW: Eight with additional backup crew on long flights

C-123X Provider

FAIRCHILD C-123 Deliveries of the Provider were begun in 1955. Rigged with skis, so that it could land and take off from snow-covered fields on rescue missions, the C-123B was used along the Distant Early Warning (DEW) line, stretching from Greenland to Alaska. Later the Provider was used in Vietnam. Engines were bolted to the wings and fuel carried in tanks underneath, making field maintenance easy.

SPECIFICATIONS (C-123K)

CONTRACTOR: Fairchild Aircraft Corporation

POWER PLANT: Two Pratt & Whitney R-2800-99W radial-piston engines, 2,300 hp each, and two General Electric J85-GE-17 booster turbojets, 2,850 hp each

DIMENSIONS: Wingspan, 110 ft; Length, 76 ft, 3 in

MAX TAKEOFF WEIGHT: 60,000 lb

MAX LEVEL SPEED AT 10,000 FT: 228 mph

RANGE WITH MAX PAYLOAD OF 15,000 LB: 1,035 miles

LOAD: 60 equipped troops or 50 stretchers, plus six ambulatories and six attendants, or miscellaneous freight

CREW: Two to four

C-130E Hercules

LOCKHEED C-130 A medium-range tactical airlift aircraft, more than 50 versions of the Hercules have been produced since the C-130 joined the U.S. Air Force in 1956. The C-130 has served, among other airlift roles, as a cargo-troop carrier, flying ambulance, in-flight refueling tanker, search and rescue plane, retriever of space hardware, hurricane hunter, drone launcher, and airborne command post. During the Vietnam War, the C-130 was converted to a side-firing gunship (AC-130A/H Spectre), primarily for night attacks against ground forces in Vietnam.

SPECIFICATIONS (C-130H)

CONTRACTOR: Lockheed-Georgia Company
POWER PLANT: Four Allison T56-A-15 turboprops, 4,300 hp each
DIMENSIONS: Wingspan, 132 ft, 7 in; Length, 97 ft, 9 in
MAX TAKEOFF WEIGHT: 155,000 lb
SPEED: 386 mph with 155,000-lb takeoff weight
RANGE: 2,500 miles with 25,000-lb cargo, or 5,200 miles with no cargo
LOAD: 47,000-lb cargo (C-130E/H), 92 troops, 64 paratroops, or 74 litter patients with two attendants
CREW: Five

C-133 Cargomaster

DOUGLAS C-133 Although the Cargomaster was not much larger in overall dimensions than the Globemaster II, the C-133 could carry twice the cargo when it arrived at Air Force bases in 1957. The C-133, a high-wing monoplane with standard tail section and paired wheel sets supported by a nose gear, was the largest production model prop-driven cargo transport in airlift history. Inter-Continental Ballistic Missiles could be loaded through its mammoth rear cargo doors.

SPECIFICATIONS (C-133B)

CONTRACTOR: Douglas Aircraft Company

POWER PLANT: Four Pratt & Whitney T34-P-9WA turboprops, 7,500 hp each

DIMENSIONS: Wingspan, 179 ft, 8 in; Length, 157 ft, 6 in

NORMAL TAKEOFF WEIGHT: 286,000 lb

MAX LEVEL SPEED AT 8,500 FT: 359 mph

RANGE WITH 51,850 LB: 4,030 miles

LOAD: 200 fully armed troops or two loadmasters and 110,000 lb of freight

CREW: Four

Camouflaged C-141B Starlifter

LOCKHEED C-141B A high-sweptwing monoplane with a T-type tail section, the Starlifter joined MAC in 1965. The Starlifter can lift combat forces over long distances, inject those forces either by airland or airdrop, resupply those employed forces and remove the sick and wounded from the war zone to advanced medical facilities. By June, 1982, all the C-141A Starlifters had been "stretched" by 23 feet, 4 inches, given in-flight refueling capability, and redesignated C-141B.

SPECIFICATIONS (C-141B)

CONTRACTOR: Lockheed-Georgia Company

POWER PLANT: Four Pratt & Whitney TF33-P-7 turbofan engines with a thrust of 21,000 lb each

DIMENSIONS: Wingspan, 159 ft, 11 in; Length, 168 ft, 4 in

MAX TAKEOFF WEIGHT: 323,100 lb

SPEED: 571 mph at 25,000 ft

RANGE: Unlimited with in-flight refueling

LOAD: 200 troops, 155 paratroops, 103 litter and 14 ambulatory patients and/or attendants, or 69,925 lb cargo

CREW: Six, including two loadmasters

135

Camouflaged C-5 Galaxy

LOCKHEED C-5 The Galaxy is the world's largest aircraft. It can carry outsize cargo at intercontinental ranges and jet speeds, yet can take off and land in relatively short distances. Like the C-141, the C-5 has a distinctive high-flying T-tail, 25-degree wing sweep, and four jet engines mounted on pylons beneath the wings. Since the first C-5 was accepted by MAC in 1969, the Galaxy's wings have been modified for greater strength and its avionic capability improved.

SPECIFICATIONS (C-5)

CONTRACTOR: Lockheed-Georgia Company
POWER PLANT: Four General Electric TF-39 turbofans with 38,800 lb thrust each
DIMENSIONS: Wingspan, 222 ft, 8 in; Length, 247 ft, 8 in
MAX TAKEOFF WEIGHT: 769,000 lb
SPEED: High cruise, 541 mph; average cruise, 517 mph
RANGE: 100,000 lb load, 4,860 miles
LOAD: Max payload weight, 242,500 lb; troops, upper aft compartment, 73
CREW: Normal, 7; minimum 4

Artist's conception of the proposed new C-17

DOUGLAS C-17 In addition to increasing MAC's strategic and tactical airlift capability, the proposed C-17 will serve as a replacement for some of the aging C-141's and C-130s in MAC's inventory. The C-17 will be able to carry outsize cargo as well as take off and land on runways as short as 3,000 feet and only 90 feet wide.

SPECIFICATIONS (C-17)

CONTRACTOR: McDonnell-Douglas Corporation

POWER PLANT: Four fully reversible Pratt & Whitney 2037 engines with 37,000 lb thrust each

DIMENSIONS: Wingspan, 165 ft; Length, 175 ft, 2 in

MAX TAKEOFF WEIGHT: 570,000 lb

SPEED: Cruise speed, 517 mph

RANGE: Fully loaded, minimum unrefueled range of 2,400 miles at 28,000 ft

LOAD: Max payload, 172,000 lb

CREW: Normal, 7; minimum 4

Index

Search and Recovery Satellite Aided Tracking program (SARSAT), 51
Search and Rescue Satellites, 51
Special Air Warfare/Air Commando forces, 97
Special Assignment Airlift Missions (SAAMS), 34, 36, 37
Special Operations Combat Control Team, 100, 102
Special Operations Forces, 25, 93–102
Suez crisis, 23
Super Jolly Green Giant. *See* HH-53

T-23, 96
Tactical Aeromedical Evacuation Subsystem, 81
Tactical Air Command, 20, 24
Tempelhof Airport (Berlin), 21
Travis Air Force Base (California), 37, 77
"Trolling for fire," 46
Troop Carrier Command, 19
Truman, Harry S., 23, 84–85
Tunner, Maj. Gen. William H., 15, 21

UH-IN Twin Huey, 90, 99

United Nations, 29, 33
"Urgent Fury." *See* Grenada

Vietnam War, 23–25, 41, 43, 49, 62, 64, 76–78, 84, 95–99, 100

Washington, George, 61
WB-29, 60
WC-130, 56–59, 64
Weather reconnaissance, 49, 55–67
Weather satellites, 62, 64
West Berlin, 20
Wingate, General Orde, 94
Women pilots, 15–16
Women's Air Force Service Pilots (WASP), 16
Women's Auxiliary Ferrying Squadron (WAFS), 15–16
World War I, aeromedical evacuation and, 71
World War II, 13, 43, 72, 105–107, 116, 117, 127
aeromedical evacuation and, 72–73

Yom Kippur War, 34

Zaïre. *See* Belgian Congo